CAMBRIDGE LIBRARY COLLECTION

Books of enduring scholarly value

Astronomy

From ancient times, humans have tried to understand the workings of the world around them. The roots of modern physical science go back to the very earliest mechanical devices such as levers and rollers, the mixing of paints and dyes, and the importance of the heavenly bodies in early religious observance and navigation. The physical sciences as we know them today began to emerge as independent academic subjects during the early modern period, in the work of Newton and other 'natural philosophers', and numerous sub-disciplines developed during the centuries that followed. This part of the Cambridge Library Collection is devoted to landmark publications in this area which will be of interest to historians of science concerned with individual scientists, particular discoveries, and advances in scientific method, or with the establishment and development of scientific institutions around the world.

Astronomical Dialogues between a Gentleman and a Lady

An Anglican clergyman and fellow of the Royal Society, John Harris (*c*.1666–1719) was an important promulgator of Newtonian science, through private teaching, public lectures and published writing. His *Lexicon Technicum* (1704) may be considered the first encyclopaedia in English. In the present work, published in 1719, Harris presents for his well-to-do readership a series of didactic conservations between a gentleman of science and an aristocratic lady. He aims to induce 'persons of birth and fortune' to dedicate some of their 'happy leisure … to the improvement of their minds', and uses quotes from poets such as Samuel Butler and John Dryden to help elucidate scientific concepts. In particular, Harris explains the use of contemporary scientific apparatus (and expensive status symbols) such as terrestrial and celestial globes. The book ends with a description of the ultimate contemporary symbol of scientific refinement: the orrery, a working model of the solar system.

Cambridge University Press has long been a pioneer in the reissuing of out-of-print titles from its own backlist, producing digital reprints of books that are still sought after by scholars and students but could not be reprinted economically using traditional technology. The Cambridge Library Collection extends this activity to a wider range of books which are still of importance to researchers and professionals, either for the source material they contain, or as landmarks in the history of their academic discipline.

Drawing from the world-renowned collections in the Cambridge University Library and other partner libraries, and guided by the advice of experts in each subject area, Cambridge University Press is using state-of-the-art scanning machines in its own Printing House to capture the content of each book selected for inclusion. The files are processed to give a consistently clear, crisp image, and the books finished to the high quality standard for which the Press is recognised around the world. The latest print-on-demand technology ensures that the books will remain available indefinitely, and that orders for single or multiple copies can quickly be supplied.

The Cambridge Library Collection brings back to life books of enduring scholarly value (including out-of-copyright works originally issued by other publishers) across a wide range of disciplines in the humanities and social sciences and in science and technology.

Astronomical Dialogues between a Gentleman and a Lady

Wherein the Doctrine of the Sphere, Uses of the Globes, and the Elements of Astronomy and Geography are Explain'd

JOHN HARRIS

CAMBRIDGE
UNIVERSITY PRESS

University Printing House, Cambridge, CB2 8BS, United Kingdom

Cambridge University Press is part of the University of Cambridge.
It furthers the University's mission by disseminating knowledge in the pursuit of education, learning and research at the highest international levels of excellence.

www.cambridge.org
Information on this title: www.cambridge.org/9781108080194

© in this compilation Cambridge University Press 2015

This edition first published 1719
This digitally printed version 2015

ISBN 978-1-108-08019-4 Paperback

This book reproduces the text of the original edition. The content and language reflect the beliefs, practices and terminology of their time, and have not been updated.

Cambridge University Press wishes to make clear that the book, unless originally published by Cambridge, is not being republished by, in association or collaboration with, or with the endorsement or approval of, the original publisher or its successors in title.

The original edition of this book contains a number of oversize plates which it has not been possible to reproduce to scale in this edition.
They can be found online at www.cambridge.org/ 9781108080194

Aſtronomical Dialogues

Between a

GENTLEMAN

AND A

LADY:

WHEREIN

The Doctrine of the SPHERE,

Uſes of the GLOBES,

And the Elements of ASTRONOMY and GEOGRAPHY are Explain'd,

In a Pleaſant, Eaſy and Familiar Way.

With a Deſcription of the famous Inſtrument, called the *ORRERY.*

By *J. H.* F. R. S.

LONDON:

Printed by *T. Wood*, for BENJ. COWSE, at the *Roſe* and *Crown* in St. *Paul's* Church-yard, 1719.

TO THE

Lady *CAIRNES*.

MADAM,

AS the Design of these Dialogues carries them naturally into the Patronage of the Fair Sex; so your own Merit, and my Duty, determine them to your Ladyship.

To you Madam! who are blest with all those Natural Graces and Genteel Accomplishments, which justly command universal Esteem; while Persons of true Taste and thorough Knowledge of Life, with Pleasure see even *those* exceeded by intellectual Beauties, and such as claim Addresses of this Nature. For what can be more engaging than to find at Lady CAIRNE's Table, the greatest Liberality and Elegance of Entertainment, outdone by improving Conversation; and the Understanding more regaled than the Senses?

But I know I must forbear; and not offend such a Modesty as your's, even with Truth: However, I can't help shewing that I am neither insensible of what all the World admires, nor ungrateful for the Obligations you have so generously conferr'd on,

MADAM,

Your Ladyship's

most humble Servant,

J. HARRIS.

THE PREFACE.

THIS *Book was most of it written a good while ago: And being supposed to be lost for some Years, was lately retrieved, and reviewed by its Author, with the Disinteressedness of a Stranger. However, I liked it so well, as to resolve upon its present Publication, with some few Emendations and Additions. Of which latter sort the Description of the famous* Orrery *of Mr.* Rowley, *is the most considerable.*

I wrote it in this diverting Way, in pursuit of a Design, *which, as I have made the general Business of my Life, so I can look back upon its Success with Pleasure,* viz. The en-

engaging Persons of Birth and Fortune in a warm Application to useful and real Learning: *To induce them to detach some of their happy Leisure from being lost by Sports, Play, or worse Avocations, and to dedicate it to the Improvement of their Minds.*

For I have often been ashamed and shocked to see, how awkwardly the few Modest have lookt, in Conversations where they could bear no part; and how insolently others have despised what they neglected to understand.

But what glorious Improvements might one expect from Persons of Fortune and Leisure, if they would addict themselves to these Things? Who can bear the expence of Good Instruments for Cælestial Observations.

For tho' there can hardly be above a Score in an Age who have pursued these Studies thoroughly: Yet such great Lengths *have been run in spite of all Disadvantages, as may*

may easily convince us, what to have hoped for, if Great Men *would now and then divert themselves this way.*

The Reader will easily see that the Conversation in these Dialogues is feigned, *and in Imitation of* Those *of the excellent Mr.* Fontenelle, On the Plurality of Worlds. *And that the* Digressions, Reflexions, Poetry *and* Turns of Wit, *are introduced to render* Those Notions *pleasing and agreeable, which perhaps without such a kind of Dress, would appear too crabbed and abstracted.*

However, I don't perplex my Fair Astronomer *with any thing but the* true System of the World: *I mislead her by no Notions of* Chrystalline Heavens, *or* Solid Orbs: *I embarrass her with no clumsey* Epicycles, *or imaginary and indeed impossible* Vortices: *But I shew her at first the Cælestial World just as it is; and teach her no* Hypotheses, *which, like some other things taught at Places of great Name,*

must

must be unlearned *again, before we can gain* True Science.

And as I think it practicable to explain and teach any Science *in this* Facetious way (Facete enim & commode dicere quid vetat ?) *so perhaps I may hereafter, if God grant me Health, Ease and Leisure, make some other Attempts of this kind. For the Lady may well be supposed, tho' the Sight of the Globes first struck her Fancy and turned her Desires this way, to have made Excursions into* other Parts of Mathematicks, *and to have discoursed with her Friend on those Subjects. And perhaps all* Those Dialogues *may not be lost, as* these *had like to have been ; but may, if these find a suitable Encouragement, be communicated also to the World.*

> *Multaq; pratera tibi possum Commemorando,*
> *Argumenta, fidem dictis conradere nostris :*
> *Verum animo satis hæc Vestigia parva sagaci*
> *Sunt ; per quæ possis cognoscere cætera tute.*
>
> Lucret Lib. I.

Aſtronomical Dialogues

BETWEEN A

GENTLEMAN and a LADY.

IT is now about ſeven Years ago, ſince I preſented the moſt Engaging Lady *M.....* with Mr. *Fontenelle*'s Book of the Plurality of Worlds: And I remember well what ſhe ſaid a few Days after.

I have look'd over your Book, Sir, ſaid ſhe, as my way is, firſt *curſorily*, and I intend to give it a *very careful ſecond Reading*; but I perceive by it, you have cut out much more Trouble for your ſelf, than perhaps you imagin'd: For I find there are many things previouſly neceſſary to the underſtanding it, which you muſt oblige me with explaining; but, continued ſhe, a Converſation of that kind with me, I doubt, will be too dull and tedious, ſince I am not bleſs'd with any of thoſe ſhining Qualifications, with which Mr. *Fontenelle* hath complimented M. *la Marquieſe*; I ſhould indeed, ſaid ſhe, except *thoſe two*, which I ſuppoſe, in Complaiſance to our Sex, he makes the Foundation of Philoſophy, *viz.* *Ignorance* and *Inquiſitiveneſs* for thoſe I'm ſure, I have in Perfection, as you have long experienced.

I need not mention the Return I made, nor how prettily ſhe changed the Diſcourſe to ſomething more general, when ſhe found I was going to ſay juſt things of her; thoſe that knew her, don't want to be reminded of the many Beauties, both of Mind and Body, which render'd Lady *M*.... one of the moſt agreeable Perſons of her Sex;, which yet were ſhe living, tho' a juſt Debt to her Merit, I muſt not have ſaid, for fear of offending her Modeſty.

All that is neceſſary to introduce what follows, is, to inform you, That ſome Years before her Death, when I went to viſit that accompliſh'd Lady at her Country Seat; I was a little ſurpriſed to find her, the next Morning after my Arrival, ſtudiouſly viewing a pair of large Globes, which ſtood in the Drawing-Room, looking into the Garden, and which I uſed to make my Place of Study.

Good

GOOD Morrow, said I, Madam, what! hath *Fontenelle* made an Astronomer of you in good earnest? Are you really contemplating the Order and Motions of the Heavenly Bodies? Or are you rather seeking on the Earthly Globe, where to make new Conquests?

The Historians foolishly represent *Alexander* the Great, as Weeping, that he could carry *his* no further than over all the World; but I'm sure, were he present now, to see you in that Posture commanding the *Globe*, and giving what Turns you please to it; that Thought of your humble Servant's would appear just enough;

Had the Pellæan *Chief thy Form but view'd,*
With far more Haste he had the World subdu'd:
Proud at thy Feet to lay the mighty Ball,
Whose Eyes were form'd to Triumph over all;
And then most justly had he Wept to see,
One World too mean an Offering for Thee!

O! Sir, said she, your Servant, I doubt you did not rest well last Night? What did your Imagination carry you into the Poetical Regions of *Fairy-Land*, that you awake with Verses in your Mouth this Morning? But to speak seriouſly, I wonder you don't bluſh to paint ſo much beyond the Life, and yet ſuppoſe the Picture to be like any one; you affect to imitate our great Painters, if we ſit to them, they make us all handſome; but they do it to ſhew themſelves, not us, and they don't care ſo much whether it be like or no, ſo it be but a fine Picture; and in this our own Vanity too often indulges them.

But pray, added ſhe, let us lay aſide all theſe Fooleries; and be ſo good as to be ſerious with me for an Hour or two: I have a great Mind to be let a little into the Knowledge of theſe Inſtruments, *the Globes*; and to know ſomething of the firſt Principles and Rudiments of *Aſtronomy*; or elſe I find I ſhall loſe half the Beauties of that very entertaining Book Mr. *Fontenelle*'s *Plurality of Worlds*, which you formerly obliged me with, as well as perhaps be led into ſome Errors by it: And don't deſpiſe and neglect me becauſe I am a Woman. I have heard you ſometimes ſay, you thought that there was no difference of Sexes in Souls; nay, that *our* Parts and Natural Capacities were often *equal*, at leaſt, if not *ſuperior*, to thoſe of Men. But perhaps there were ſome particular Reaſons for your ſaying ſo then, which now altering or ceaſing,

ceaſing, your Judgment and Opinion may have done ſo too.

I was going to aſſure her, that I was ſtill of the ſame Sentiments, when putting on a forbidding Look, with a ſerious Countenance ſhe proceeded thus :

Globes.

Theſe Globes, Sir, came too late to accompany a Relation of mine to *India*, his Ship having ſailed before they were finiſhed, which is the Reaſon you ſee them here ; and I have ordered them to be ſet out this Morning, and ſhall do ſo from Day to Day, tho' without obliging you to what *Fontenelle* had with the *French* Lady, an *entire Week's Conference*. But I have a great Mind to learn, from my Friend, ſomething of the Nature and Uſe of them ; for they appear to be made and finiſhed up with that Curioſity and Care, that ſure ſome very uſeful Knowledge is to be learnt from them, and is it not barbarous in you Men to confine it all to your ſelves ?

MADAM, ſaid I, you will give me a new Riſe to value any thing that I underſtand ; if I can render it acceptable to you.

WELL then, Sir, said she, all Compliments apart, both to your self and me, pray let us go to our Business, the Tea won't be ready this Hour, and there is a little too much Dew for us to take a Walk in the Garden. Let me understand then, first the Difference between these two Globes, and why one hath the Cities, Countries, and Places of the Earth drawn on it, like a Map; and the other *Circles* and *Stars*, and these odd uncouth Figures of Beasts, Birds and Fishes: Pray why do they turn round? What doth this Brass Hoop signify in which they hang? For I perceive that it also hath Numbers engrav'd upon it: And what doth this *broad wooden thing* serve for, that hath the Days of the Month and other Letters, as well as Figures, pasted upon it?

I am glad said I, Madam, by the warm Manner of your Enquiry, to find that you are in earnest; and I have often wished that the same Curiosity and Love of Knowledge would inspire more of the fair Sex, for it would mightily enlarge their Empire and Power over ours, by endowing them with more real and lasting Beauties, such as would improve with Time, and strengthen even in Age itself. But

But as to your preſent Queſtions, Madam, I will give you the moſt Satisfactory Returns I can.

And firſt, Madam, it will be neceſſary to acquaint you with the Meaning of the Word *Globe*; and what the Properties, in general, of ſuch a Figure or Body, are.

Globe what. Your Ladyſhip is to underſtand then, that a Globe is a round Body of ſuch a Nature, that every Part of its Surface or Out-ſide, is at an equal Diſtance from one Point within it, which is called the Center. This Body alſo is ſometimes named a *Sphere*, with regard to Aſtronomical Speculations; and this Science which you are now inquiring into, is hence called *The Doctrine of the Sphere*. *Sphere.*

I THINK I underſtand you; ſaid ſhe, the Figure of a Globe is not flattiſh like that of a Cheeſe or a common Ninepin-Bowl; but rather like a Boy's Marble, or a Bullet caſt in a Mould.

EXACTLY right, Madam, ſaid I, and further you are to know, that a ſtrait Line ſuppoſed to be drawn thro' the Center of this Globe any where, from one oppoſite Point of the Surface to the other, is called a *Diameter*. *Diameter*

I THANK

I THANK you, said she, for that Explication, Sir, I have often met with the Word, but never knew *fully* what Diameter signified before: But now I know what the ingenious Mr. *Butler* meant when speaking of the Moon, he saith, that *Sydrophil* knew

What her Diameter to an Inch is,
And prov'd she was not made of green Cheese.

And now I know what the Plummer meant the other Day, when he talk'd of a Pipe of Lead of such a Diameter; I now know the Meaning of *Diametrically opposite*, &c. But, pray, Sir, go on.

YOU will next see easily, Madam, said I, that if a Globe were at Liberty, and any Power or Force at hand to move it, it would easily turn or roll round any one of its Diameters, as this Globe doth round this *Wire*; which particular *Diameter*, is called therefore its *Axis*; as being the *Axle-tree* on which it turns. But tho' this be true of the Nature of a Globe in general, yet the *Axis*, as we call it, of the Earth and Heavens, by the Will of our All-wise Creator, is one *fixed and determinate Line*; and about this the

Axis.

fixed

fixed Stars are uſually ſuppoſed to revolve, without ever changing their Diſtance, or deviating from one another or from it.

I AM mightily pleaſed, returns ſhe, with the Nature of theſe Globes, becauſe they are unbiaſſed and indifferent, as to this or that particular Way of Turning; and I fancy it to be a good Emblem of the Freedom of our Minds in the State of Innocence, when they firſt came out of Nature's Hands; they were then perfectly at Liberty to move any way, which they lik'd beſt; and I dare ſay, that all the wrong Biaſſes and particular Turns that we find in any of them, are owing to the Weight or *Power*, as you call it, of our own corrupt Affections.

YOU moralize excellently well, ſaid I, Madam, and are very juſt in your Notions of the Deity.

But ſhe went on, and ſaid; Yet I think we might be glad to receive from the firſt Mover and Author of all Things, ſuch a *determinate Way of moving*, as you ſay God hath given to the Heavens and the Earth; for our own whimſical Motions, Turnings and Shiftings, ſeem to be as unaccountable as they are various.

BUT

Motion of the Heavens. BUT pray, ſaid ſhe, let me underſtand what you ſay as to the preſent Point a little further; Do the Heavens and the Earth all really move round about one *Axis*, as theſe two Globes do round theirs? And are the *Poles* thus beautifully deſcribed by Mr. *Dryden*, the two Ends of this Axis?

Poles.

Two Poles turn round the Globe, one ſeem to riſe
O'er Scythian *Hills, and one in* Lybian *Skies*;
The firſt ſublime in Heav'n, the laſt is whirl'd
Below the Regions of the nether World;
Around our Poles the ſpiry Dragon glides,
And like a wandring Stream the Bears *divides,*
The Leſs *and* Greater, *who by Fate's Decrees*
Abhor to dive beneath the Southern Seas;
There, as they ſay, perpetual Night is found,
In Silence brooding on th'unhappy Ground:
Or when Aurora *leaves our Northern Sphere,*
She lights the downward Heav'n, and riſes there,
And when on us ſhe breaths the living Light,
Red Veſper *kindles there the Tapers of the Night.*

DRYDEN's *Virgil.*

Shall I ever come to know what theſe *Poles*, and *Dragons*, and *Bears*, mean?

VERY

VERY eaſily, Madam, ſaid I, and you will find that the *Motion of the Earth alone round its Axis* will ſufficiently account for all the reſt; for theſe *fixed Stars* don't in Reality move at all, but only appear ſo to do. And you muſt know, that there is one Star, or a Point very near it, towards which this *Pole*, or End of the Earth's *Axis*, (which is called the *North-Pole*) doth always point: This is the Star here on this Celeſtial Globe, and if it be fair, and the Sky clear, in the Evening, I will ſhew it you in the Heavens: 'Tis ſaid, by Aſtronomers, to be in the Tip of the Tail of the *Little Bear*, a *Conſtellation of Stars* ſo called; you ſee there are ſeven of theſe Stars in all, placed on the Globe within the *Picture* or *Figure of a Bear*: The Reaſon of the Figure I will tell you hereafter.

Motion of the Earth.

Pole Star.

PRAY, ſaid ſhe, good Sir, don't take it amiſs if I interrupt you with one Queſtion: Is this *Tip* of the Bears Tail, *that celebrated Tip* of *Cardan* the Conjurer; who, as *Butler* ſaith,

Firmly believ'd great States depend,
Upon the Tip of th' Bears Tail's End,

That

That as ſhe whisk'd it tow'rds the Sun,
Strow'd mighty Empires up and down.

THE very ſame, Madam, ſaid I.

Go on then, ſaid ſhe.

THIS Star here by the Wire, Madam, ſaid I, we call the *Pole Star*, and the Point near it, thro' which the Wire runs, the *North Pole* of the World. And let the Earth be where it will, in its Annual Courſe round the Sun, this *North Point* on the Earth, and here placed on the Globe, will always be either exactly or nearly under that *North Pole* Star or Point, in the Heavens. But of this more when I ſhall further explain to you the Motions of the Earth; and this Poſition of the Earth's Axis is ſo firmly fixed and determined by the Author of Nature, that from it there hath never yet been obſerved any conſiderable Variation.

PRAY, Sir, ſaid ſhe, proceed: When I come to look over *Fontenelle* again, I perceive I ſhall underſtand him and you much better.

MADAM, ſaid I, the outward Figures of theſe two Globes you ſee are nearly alike; but tho' they are hung alſo, and fitted

fitted up alike, yet they are almost as different from one another in their Natures and Properties, as are the different Regions that they represent.

Terrestrial Globe. This Globe which is designed to shew the Face of the *Earth*; and which therefore is called the *Terrestrial Globe*, is truly and properly a Representation of it, round or spherical as that nearly is, and it hath the Sea and the Land, with all the Regions, Countries, Nations, Islands and Cities drawn upon it; just in that Order and Figure, that they are, in Reality, on the Face of the Earth itself; and it is, if carefully drawn, a true Map, or Description, of what is usually called *The World:* whereas all those flat Maps and Charts, which you see drawn upon Paper, cannot be accurately so, tho' they are exact enough for common Use.

THAT Word *World*, said she, I can't get over without reflecting, what weak, vain, and silly Mortals we are: We too often take this *poor Spot of Earth* to be the only *World* worth inquiring after; and so we can but acquire a little of its Dirt, we neglect all Care for an Eternal Mansion in the Heavens. And further, I have no Patience with *Ptolomy*, I think they call him, and his Astronomers, that will needs have the mighty

mighty Sun, and all that infinite Orb of fixed Stars, to be made only for the ſake of *this little dirty Planet*, as I remember ſomebody calls it; and to have no other Uſe nor End, but only to dance round it, which yet, as I have heard, is a meer Point, and ſcarce viſible to an Eye placed in ſome of the other Planets.

But to go on with my Leſſon: Good Sir, ſaid ſhe, is the Figure of the Earth thus really round? and have you any good Reaſons to make you think ſo? For I muſt own I had not till now a Notion of its being round like a Ball; I took it rather to be round in Compaſs like a Diſh or Plate.

Rotundity of the Earth.

VERY many and ſubſtantial ones, Madam, ſaid I, and you will be fully convinced by them, when they occur to your Reading hereafter, if you proceed on in that Way you are now going: But, however, the Sun ſhining ſo bright into this Room, will furniſh me *now* with one Argument to make that Notion plain to you. You ſee, Madam, when I hold any ſolid Body in this Light of the Sun, its Shadow will be nearly like the Shape and Form of that of the Body; when I hold this Book in the Light, its Shadow will be be square at the Sides, as the Book is; but when I hold this Orange in the same Light, the Shadow, you see, hath a round Edge; and therefore since in the Eclipses of the Moon, the Shadow of the Earth, which you know, Madam, occasions the Moon's being covered with Darkness, appearing always exactly round or circular, we justly conclude that the Figure of the Earth is round or spherical too, or else the Termination or Out-Line of its Shadow could never be always in a Circular Form.

I THANK you for this easy and natural Explication, said the Lady, which I think I comprehend; and I am beholding to the Sun, that great Fountain of Light, or rather to Him that made it, for being now instrumental to dispel the Darkness I had in my Mind before about this Affair; however, being no *Persian*, I shall not worship the Sun for it. But pray, Sir, go on with an Explication of the other Globe.

Celestial Globe.

THAT, Madam, is called the *Celestial one*, said I, because 'tis designed for a *Representation of the Firmament, and the Concave Arch of the Heavens*; and indeed it doth well enough exhibit to us the *fixed Stars*, and the Tracks or Circles of the

the Sun and Planets apparent Motions, if you get a right Notion of it, as this Figure, which we call an Armillary Sphere, will I think help you to obtain: In order to which you muſt now imagine your Eye placed within at the Center of the Globe, or on the little Ball there in the Figure which repreſents the Earth; and that the Spherical Surface of it, on which you ſee the Stars there painted and gilded were tranſparent like Glaſs; ſo that you could actually ſee thro' it, not only all the Circles drawn upon it, but alſo all the Stars above in the Heavens, as they really appear there in a bright Night. And if you imagine further, Madam, that ſtrait Lines were drawn from every Star in the Firmament to your Eye ſo placed, as before, in the Center of this Globe, thoſe Lines would paſs thro' and cut the Spherical Surface of the Globe in proper Points to paint, or to place the Pictures of the Stars upon.

Vid Fig. 1.

I THINK, I conceive you right, ſaid the Lady, ſo that if there were Holes in the Surface of this Globe in thoſe Places where theſe Stars are painted upon it, and that my Eye were within at the Center, and the Globe turn'd, ſo as to conform itſelf to the preſent Poſition of the Heavens

Fig. I.

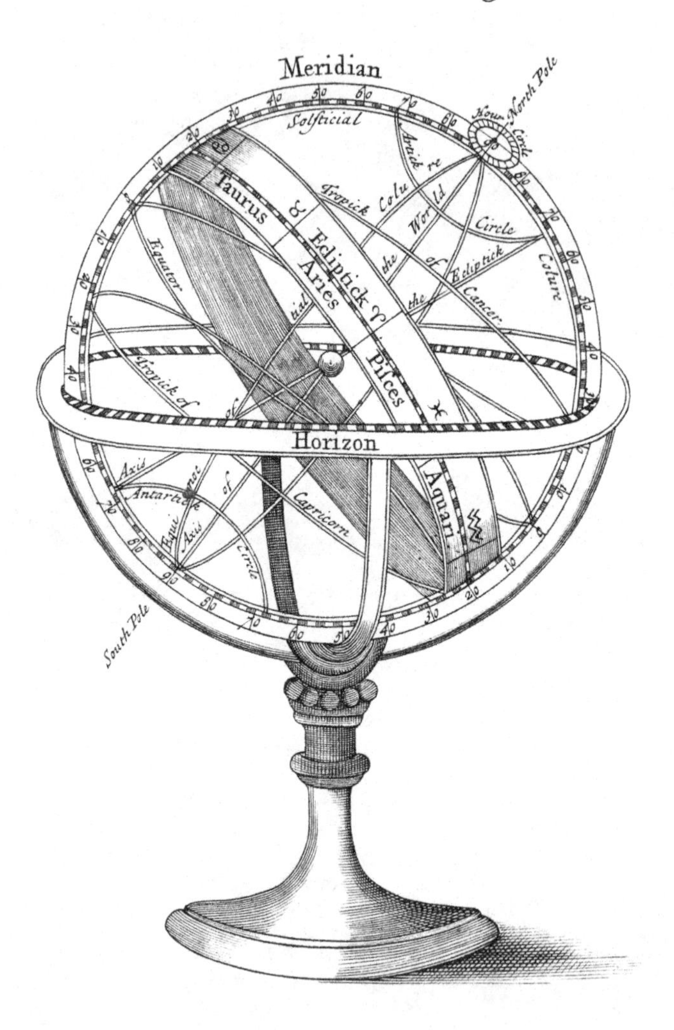
Meridian
North Pole
Hour Circle
Solsticial
Artick
Colure
Tropick
the World
Circle
the Ecliptick
Cancer
Colure
Taurus
Ecliptick
Aries
Pisces
Equator
Tropick of
Horizon
Axis
Antartick
Axis
Capricorn
Circle
Aquari.
South Pole

Heavens above; I ſhould ſee every Star there thro' its correſponding Hole in the Globe.

YOU are perfectly right, Madam, ſaid I, and *Ptolomy* himſelf, could not have expreſſed it better. And juſt in that Central Point (and juſt ſuch a Point as that is it) do Aſtronomers of his Sect ſuppoſe the Earth to be placed, as you ſee in the Figure, in the middle of the Sphere of the fixed Stars, which ſeem to revolve round about it, once in 24 Hours, becauſe the Earth doth turn round her own Axis, tho' a contrary Way, in the ſame Time.

Vid. *Fig.* I.

OF this, replied *Clarella*, I have gotten a tollerable Notion from what you ſaid before, and from the *French* Author: But, pray, let us now go on with our Globes here; What is the Meaning of this *broad Wooden Circle* placed round each of them, and what is *it* called?

MADAM, ſaid I, it is called *The Horizon*; which is a *Greek* Word that ſignifies a *Limiter* or *Determiner*. And to conceive it right, imagine your ſelf placed, as before, on this poor little Earth, within that immenſe Celeſtial

Globe; which you are to ſuppoſe now to be millions of Millions of times greater than it really appears to be: Then you know, if you look round you on the Earth, its Surface will extend every way from your Eye, like a vaſt Plain; which will be under your Feet, and to which your Body will be perpendicular or upright: this Plain ſtretching all round you every way as far as your Eyes can ſee, in a flat open Country where no Hills interpoſe: Or on the Surface of the Sea, will ſeem to divide or cut the Concave Orb of the Stars, or the Sky, into two Parts (which they call *Hemiſpheres*; the one ſeemingly above this Plain; which therefore they call the *Upper*, and the other apparently below it: Which therefore they call *the Lower Hemiſphere*. Such a Plain as this is call'd *the Horizon*: And if it be really that which any one's particular Eye makes upon any occaſional View, 'tis call'd *the Senſible Horizon*: But if you imagine this Plain, as you may eaſily do, to paſs through the very Center of the Earth on the Surface of which you then ſtand, 'tis called *the Real or Rational Horizon*; becauſe that doth really or actually divide the Starry Regions into two equal Hemiſpheres; and both theſe Horizons are well enough repreſented by that wooden Circle,

Hemiſpheres.

Horizon.

cle, which you now lay your fair Hand upon.

I HOPE I take you right, ſaid ſhe; and now begin to underſtand better the Meaning of many Expreſſions which have often occurr'd to me before, but with leſs Light. But why do you ſo cautiouſly uſe the words *apparently above and below?*

BECAUSE, ſaid I, Madam, there is in reality no ſuch thing as any Difference between *above and below:* The Heavens are every where *above* or *without* what they contain; but we, taking our Ideas of things from ourſelves, do agree to call that *above* or *uppermoſt* which is over our Heads, and that *below*, which is *beneath* us, or down under our Feet: And therefore as we call that *Concave* Half of the Region of the Fixed Stars, which we ſee above our Horizon, the *Upper Hemiſphere*; ſo the other Half takes the Name of the *Lower Hemiſphere.*

I AM mightily pleaſed, ſaid the Lady, with theſe Celeſtial Beings that are ſo perfectly above all the poor Trifles of *Place* and *Station*; with which we Mortals make ſuch a buſtle here below: Eſpecially thoſe of our Sex; as I will honeſtly own to you, now you are my Maſter and Teacher; for as *Butler* hath juſtly obſerv'd;

To us the Joys of Place and Birth
Are the chief Paradiſe on Earth:
A Privilege ſo ſacred held
That none will to their Mothers yield,
But rather than not go before,
Will forfeit Heaven at the Door.

But let us go on. I perceive, ſaid the Lady, that theſe Horizons will always vary as we ſhift the place of our View.

YES, Madam, ſaid I, and ſo will the Hemiſpheres too that they determine.

AND yet, ſaid ſhe, we are often ſo vain as to take our little narrow View or Horizon for the Bounds of all that is to be ſeen; and judge, that what is not within our Hemiſphere, to be either nothing at all, or at leaſt not worth our knowing or enquiring after; for we are always ſo vain as to deſpiſe what we do not underſtand. But I interrupt you with my impertinent Reflections; pray, Sir, go on.

I BEG you to take notice farther, ſaid I, Madam, that when the Sun, or any Star or Planet, appears at the Eaſtern Edge of our

our Horizon, we ſay it is *Riſing* ; and when it is got quite above it, we ſay it is *Riſen*, or is Up. On the contrary if it appear towards the Weſtern Edge of it, we ſay it is *Setting* ; and when it is gotten below it, we ſay it is *Set*. And this Riſing and Setting always reſpects the ſenſible, and not the Real Horizon.

BUT what is the meaning of theſe Circles, demands ſhe, which I ſee drawn here upon the Board of the Horizon, and on both Globes alike?

Sea-Compaſs.

THE outermoſt of them, Madam, ſaid I, repreſents the *Points of the Compaſs*, as they are called by our Seamen; who make uſe of an Inſtrument called *the Compaſs*, to ſteer their Ships by at Sea.

PRAY let me know a little more of that matter, ſaid ſhe, for 'tis a Thing I have heard much talk of.

YOU have ſeen, no doubt, Madam, ſaid I, a *Loadſtone*; and know that it hath that wonderful Virtue, among others as ſtrange, that if a Needle or long Iron-Wire be drawn rightly over it, that Needle will ever after that, when at liberty, *point*, as they call it, due North and South.

You are now, ſaid ſhe, ſo very good, that I think I muſt feed your Vanity, by owning, that I was once much pleaſed with ſome Verſes of yours occaſionally given me; but am more ſo now, becauſe I underſtand them better; after you had talked in your uſual way of Love and Conſtancy and I know not what; you thus, as I remember, concluded,

So when the Needle hath been once drawn o'er
The Loadſtone's Poles, and felt its wondrous Power,
'Twill e'en in Abſence keep its Truth and Worth,
And always point *tow'rds its beloved North:*
But when it once the Magnet's Preſence gains,
With Joy it trembles and the dear Object joyns.

Madam, ſaid I, you do Me and my Trifles a great deal of Honour ———

Hush! ſaid ſhe, not a word! I won't now allow you one Syllable of Trifling; be quiet and go on with your Lecture.

Please to let me inform you then, Madam, ſaid I, that ſuch a Wire as this, ſo *touch'd*, as they call it, or directed by the Power of the *Magnet*, or Loadſtone, they put into a round piece of Paſteboard, on which they draw a Circle; dividing it as this on the wooden Horizon of the Globe

Globe is, firſt into four Quarters, for Eaſt, Weſt, North and South, placing the Point of North over that End of the Wire which will point that way; then they divide each Quarter into Halves; and by that means they make in the whole 32 Diviſions, which they call *Points*; and which are there and here expreſſed by the Initial Letters of their Names after this manner: [*See Fig.* II.] And therefore the Uſe of that Circle on the Horizon of the Globes is to ſhew, on *what Point of the Compaſs* the *Sun*, or any *Star* or *Planet* apparently *Riſes* or *Sets*; as I ſhall ſhew you more fully hereafter.

WELL! ſaith ſhe, I fancy my ſelf half a Sailor already; but for all that I muſt confeſs ingenuouſly to you, that I don't know how to find the Points of Eaſt, Weſt, North and South in the Heavens, or on the Earth, unleſs I ſee a *Church*, which, they ſay, uſually ſtands Eaſt and Weſt.

MADAM, ſaid I, that is eaſily known, by the Noon-day or Meridian Sun; for the Sun at Twelve a Clock being always full South, when you turn your Face towards it, the North will be on your Back, the Eaſt on your Left, and the Weſt on your Right Hand.

THAT'S

THAT's true, ſaid ſhe; but as obvious as this Obſervation is, I never made it before. And really the Education of us Women, is ſo ſilly and crampt, that, generally ſpeaking, we are never taught, nor innured to think of any thing out of the common Way, and beyond the Legend of the Nurſery: Nothing but our *Work*, a little *Houſwifery*, and a great deal of *Goſſiping*.

Calendar

But pray let us go on: The next Circle I perceive is only an Almanack, with both our Own, and the Foreign or New Stile, or way of accounting Time: But pray, Sir, of what Uſe is this innermoſt Circle, and how is it divided?

Diviſions of a Circle.

MADAM, ſaid I, all Circles on the Globes are ſuppoſed to be divided into 360 equal Parts, which they call *Degrees*, and each Degree into 60 leſſer Parts, which they call *Minutes*, and ſo on, by a Sub-diviſion by 60 ſtill, as far as you pleaſe. This Circle is deſign'd to ſhew us what we call the *Sun's Place* for every Day in the Year; and therfore is divided actually into 12 parts, which are diſtinguiſh'd here, you ſee, by theſe Pictures of 12 Eminent Conſtellations, or Parcels of Stars; and which, becauſe they do *ſign*

or

or mark out a *particular Place* in the Heavens, where the Sun is, or appears to be, every Month, have been called the *Twelve Signs* of the Zodiac: And each of these *Signs* is divided into 30 equal Parts or Degrees, which makes up the whole 360.

Signs of the Zodiac.

HOLD a little, Sir, said the Lady, for I have now so many things to ask you that I know not where to begin ——

MADAM, said I, all the Affair of the Zodiac, of the 12 Signs, and of the Sun's apparent Yearly Motion through them, I will fully explain to you hereafter: And all you need know now is, That it is the Use of this Circle to shew you in what Degree of it, or in *what Place* or Part of any of the 12 *Signs*, in which the Sun is supposed to be at Noon, answers to each particular Day of the Month: As for Instance; You see this Day, *May* the 20th, is placed in the Calendar, just against the first Degree of (♊) *Gemini*, and therefore that is the *Sun's Place for this Day*.

SINCE I must wait, said she, I will be patient, and be content to be taught in your own Way; but I will never forgive you if you don't tell me, *just now*, why 360 was *only* pitch'd upon for the Number of Divisions,

Divisions, or, as you call them, *Degrees of your Circles*; and why any other Number would not have done as well?

MADAM, said I, any other *greater Number* that could have been broken into *Parts without Fractions* would have done better. But they had a particular Reason to pitch upon this of 360, which yet I beg you will excuse me from telling you now, because it will be much more usefully explain'd hereafter, and save a great many Digressions at present.

WELL! said she, I'm sure you keep me out of this only to mortify me, and to try my Patience; but that I may not tire *yours*, I submit.

YOU are so moderate and easy in your Desires, Madam, reply'd I, that I will now go out of the common Method, and explain all that matter to you immediately.

Sun's Motion.

The Ancient Astronomers observed of the Sun, that besides his apparent Motion round the Earth in 24 Hours, by which he made, as they supposed, Day and Night; the former when he was above, the latter when he was below the Horizon of any place; which Daily or *Diurnal*

nal Motion (by the by) they supposed to be always made either in this very Equinoctial Circle, or in some other lesser ones parallel to it, or equally distant from it: These Parallel Circles also they supposed to be, in the Summer Half-year on the *North side*, and in the Winter, on the *South side* of the Equinoctial. And they took notice, Madam, that besides this Diurnal Motion (which appear'd to be circular) the Sun had also in appearance a progressive one, *forward* on in another circular Track in the Heavens; which, because they found that when ever the Moon came into the very same Circle, there would be an Eclipse of either *Her*, or of the *Sun*, they call'd the *Ecliptick*. This is the Circle here on the Globe, which lies oblique to, or *askew*, and cuts or crosses this other, which is drawn exactly in the middle between the Poles, and is call'd the *Equinoctial* or *Equator*: This Ecliptick Circle also, because they perceived that the Sun never deviated from it in his Annual Motion towards either Pole North or South, they called the *Way of the Sun*: And they found that in the Time of our Common Year, he would appear to go quite round, or pass successively through all the Parts of this Circle.

But, said the Lady, how could they determine that? For when the Sun was above the Horizon, no Stars at all could be seen, to distinguish his Place or Situation by.

Your Objection is just, said I, Madam, if you consider the thing *after* the Sun was actually Risen, and just *before* his Setting: But they took notice of those Stars which were at or near the Edge of the Horizon *before* his Rise, and such as were there *after* his Setting; and found that the Sun would not continue to rise and set always at the same distance from the same Stars; but if, for instance, on *March* the 10th, he would rise and set near these Stars which you see here placed on the Globe within this Constellation called *Aries*, about a Month after they found that he would rise and set with those in *Taurus*, which lie a 12th part of the whole Circle more this way, or forwards on, as the Numbers shew, to the Eastward; and after this manner the Sun proceeding still forward every Day, they found that at the end of 12 Months he would seem to have gone entirely round in this Circle, and to rise and set successively with or under all the Fixed Stars, which

which are in or near this Circular Track called the Ecliptic.

BUT, pray, Sir, ſaid ſhe, what do you mean by *under* the fixed Stars? Why, don't the Sun move *in among* them, and along with them?

NO, by no means, Madam, ſaid I, the fixed Stars are probably farther, a long way, from the Sun, than that mighty Luminary is from us; and the Meaning of the *Sun*'s *Place*, or his being in ſuch a Sign, is only his being for ſuch a Time under that Star or Conſtellation, or between that and our Eyes; ſo that if a Right Line were drawn from that Star to your Eye, it would paſs thro' the Center of the Sun.

Sun's Place.

I BEGIN, ſaid the Lady, I think, to comprehend this a little better than I did; but, pray, Sir, what is the meaning of the Word *Zodiac*, which you uſed a while ago, when you began to talk about the Sun's Motion?

Zodiac.

THE ancient Aſtronomers, Madam, ſaid I, to diſtinguiſh theſe Conſtellations, or *Setts of Stars*, under which the Sun conſtantly appeared to move in his Annual

nual Course, gave them particular Names: The first they called *Aries*, or the Ram; the second *Taurus*, or the *Bull*, &c. and because these Names were mostly taken from *Animals*, or living Creatures, they called it the *Zodiack*; which is a *Greek* Word expressing such a Collection.

WELL, said she, as for your *Greek*, I know nothing of the matter, but now I begin to find out the Justness of those Lines, in *Hudibras*; wherein he describes *Sydrophil*'s Surprise at the Discovery of his new Star, occasion'd by a Lanthorn at the Tail of a School-Boy's Kite:

'Tis not among that mighty Scrowl,
Of Birds, and Beasts, and Fish, and Fowl,
With which like Indian *Plantations*
The Learned Stock the Constellations.

And these, I suppose are the Pictures, continu'd she, of those *animated Stars*, or rather, as *Butler* hath it in the same Place, the *Signs of Those:*

Nor those that drawn from Signs *have been,*
The Houses *where the Planets inn.*

MIGHTY well remember'd, said I, Madam, you see at once why the Astronomers

nomers call them the *Twelve Signs*, becauſe, as I ſaid before, they *ſign* or mark out the Place of the Sun in the Heavens; and alſo why the Aſtrologers called them *Houſes*, becauſe they aſſigned them as Dwellings or Places of Abode for the Planets:

O! ſaid ſhe, now you talk of Aſtrology, I muſt ask you a few Queſtions about *that* either now or ſome other Time; for I long to know whether there be any thing in that Art or no; for I think I have heard you throw out ſome ſuſpicious Words about it.

MADAM, ſaid I, if you pleaſe to go on with your Aſtronomy, you will ſoon know enough to deſpiſe that vain and fooliſh Cheat, as a thing perfectly beneath your Enquiry into.

VERY well, ſaid ſhe, and ſo if I will be an Aſtronomer, it ſeems, I muſt at once bid adieu to that darling Pleaſure of our Sex, Curioſity, and the Deſire of knowing our Fortunes; this is very hard, and you are really, Sir, a very bad *Woman*'s-*Man*; you have Philoſophiſed me out of many a fair Pleaſure already; Cenſure, Satyr and Goſſipping are almoſt gone; and muſt dear *Inquiſitiveneſs* follow them too?

too? It ſhall never be, let it be never ſo ſilly; I remember what *Butler* ſaith:

Doubtleſs the Pleaſure is as great,
Of being cheated, as to cheat;
As thoſe receive the moſt Delight
Who leaſt perceive a Jugler's Slight;
And ſtill the leſs *they underſtand,*
The more *admire the Slight of Hand.*

but I ha'n't Time to quarrel with you, and to diſpute it out with you now; pray, therefore, Sir, go on, about the Sun's Motion, a little farther.

YOU muſt know then, Madam, ſaid I, that theſe venerable Star-Gazers, finding the Sun apparently to run thro' this Zodiac, in twelve Months, or a Year's Time, aſſigned one part of the Circle to a Day's Motion; and becauſe there are but a few more than 360 Days in a Year, they ſuppoſed this Circle of the Sun's Annual Motion, to be divided into 360 equal Parts, which they called *Degrees*, as I told you before; and hence all Circles on the Globes came to be divided after the ſame manner.

I thank you, Sir, ſaid ſhe, now this Matter begins to clear up to me; have you

you any thing more to teach me about this Circle?

ONLY the Explanation of a few Terms, or Words, which you will find uſed about it, ſaid I, Madam: For you muſt know, that the Aſtronomers call the Diſtance of the *Sun*'s *Place* at any time of the Year, from the Beginning of *Aries* here, which you ſee is placed at the Eaſtern Point, where this Circle of the *Ecliptick*, and that of the *Equinoctial* croſs one another, they call that Diſtance, I ſay, his *Longitude*; and tho' the Sun himſelf apparently moves always in one Circle, exactly in the middle of the *Zodiac*, that is in the *Ecliptic*, yet the Moon, and the other Planets, do not, but ſometimes are 5 or 6 Degrees to the North, and at others, as far to the South of this Circle; and this Deviation or Diſtance they call their *Latitude*; and you ſhall be ſhewn hereafter how to meaſure it; and the ſame Word is uſed alſo, with Reference to thoſe fixed Stars which are not in the *Ecliptick*, but are diſtant from it, any Way, towards either of its Poles; for the Diſtance of a fixed Star from the *Ecliptic*, is alſo called its *Latitude*.

Sun's Longitude.

Planets Latitude

BUT, pray, Sir, ſaid ſhe, is not the Word *Latitude* uſed alſo with Reference to the *Terreſtrial Globe*? Surely I have heard my Brother ſpeak of *Peking* in *China*'s lying in ſuch a Latitude; of the Latitude of *London*, and of his Ship being harraſſed by a Storm, in ſuch a *Latitude*; but I muſt own I never knew the, Meaning of it: Am I Aſtronomer enough to be taught that now?

YES, Madam, ſaid I, and you will very eaſily comprehend it: Pleaſe to turn your Eyes to this Terreſtrial Globe; this Circle which lies exactly in the middle, between the two Poles of the Earth, is here called the *Equator*, and by the Sailors *the Line*; all Places which lie under it, or which have the Equinoctial in the Heavens, paſſing over their Heads, are ſaid to have no *Latitude*; but all *other Places* that lie at any Diſtance from it, either North or South, are accordingly ſaid to have North or South Latitude: And its Quantity is known by turning the Globe about till the Place come to this *Brazen Circle* in which the Globe hangs, and there *the Place* will ſhew its own *Latitude*, in Degrees upon that Circle: Thus, you ſee, Madam, when

Equator.

Latitude of Places.

when I bring *London* to this Braſs Circle, it appears to lie on the North Side of the Equator, in 51 $\frac{1}{2}$ Degrees diſtant from it.

MIGHTY well, Sir, ſaid ſhe, I now conceive what *paſſing* or *croſſing the Line is*, which I have heard the Sailors make ſuch a Fuſs about; and I have read of ſtrange Ceremonies and Duckings, which they make young Navigators undergo, at the firſt Time of their *croſſing the Equator*: I perceive now, alſo, the true Meaning of ſeveral Allegorical Expreſſions, which, no doubt, are taken from hence, ſuch as being a *Latudinarian in Notions*, &c. But pray, Sir, let us go on; now you mention that *Braſs Hoop*, in which the Globes hang and turn round, pray let me know its Name and Uſe?

Meridian.

THAT Brazen Circle, Madam, ſaid I, is called the *Meridian*; and 'tis a greater Circle of the Sphere, which is ſuppoſed to paſs thro' the *Zenith* and *Nadir* of any particular Place, thro' the North and South Points of its Horizon, and thro' the Poles of the World.

I SEE, ſaid ſhe, the latter part of what you ſay; but pray, what do you mean by the Terms *Zenith* and *Nadir*, the former of which Words I have often met with in Books, but never knew the Meaning of it.

'TIS an *Arabick* Word, ſaid I, Madam, and ſignifies that Point in the Heavens that is directly over your Head, as *Nadir* doth the oppoſite one in the lower Hemiſphere, at the oppoſite End of a Diameter of the Earth : And this Brazen Circle is called the *Meridian*, becauſe, whenever the Sun comes to the Meridian of any Place on the Earth, in his daily Courſe, 'tis then, what the *Latins* called *Meridies*, i. e. *Mid Day*, or exactly *Noon* there.

O! Sir, ſaid ſhe, this *Aſtronomy* is mighty inſtructive; I now underſtand the juſt Meaning of ſuch Expreſſions, as theſe,

There Vice did in its Zenith reign,
Our bright Meridian Sun decline, &c.

But pray let me know the Uſe of this Circle here on the Globes.

I ſhew'd you juſt now, ſaid I, Madam, That on the *Terreſtrial Globe* it ſhewed the *Latitude* of all Places, which, by being brought ſucceſſively to it, as the Globe

Globe turns round its Axis, do each receive it for their own Meridian, for 'tis all one as if a different Meridian had been actually drawn on the Globe thro' every Place.

No doubt on't, ſaid ſhe, for 'tis the ſame thing, *as to meeting*, whether the Mountain walks to *Mahomet*, or He ſtalk to the Mountain: But methinks this *Earthly Meridian* is either very lazy, or elſe takes great State upon him, that all Places muſt come to him, while he ſtands and ſtruts here, and won't ſtir the leaſt Step towards them. —— Have you any thing more to tell me about this *Man of Braſs*; *Spenſer* did wiſely to make his Man *Talus* of Iron, that was to be *Arthegall*'s Page and to bear ſo buſy and active a Part in his Story.

MADAM, ſaid I, this Braſs Meridian ſerves alſo, by its moving thus, round, North or South, in this perpendicular Situation to the Horizon, to elevate or raiſe the Pole of any Place as much above its Horizon in Degrees, as is the Latitude of that Place, or its Diſtance from the Equator, and then that particular Place will be brought to lye in the Zenith, or uppermoſt Point of the Globe.

Height of the Pole.

PRAY explain this by an Instance, said she.

I SHEWED you just now, said I, that the *Latitude* of *London* is found by the Help of this Meridian to be $51^\circ \frac{1}{2}$; raise therefore the North Pole so, that the Northern Edge of the Horizon cut $51^\circ\ 30'$ of this Brass Meridian, reckoning from the Pole, and then *London* will be in the Zenith Point of the Globe.

I SEE it is, said the Lady, and I believe I see also the Reason why it must be so; for it is just as far (*viz.* 90°) from the Equator to the Pole, as from the Zenith to the Horizon; so that taking away the middle Part, which is common to both, the *Latitude* of any Place, and the *Height of the Pole* above its Horizon are all one in Quantity; and so I suppose 'tis called the *Height of the Pole*, because the *Pole Star*, which is near the Polar Point (as I think you told me) will appear, in the Night, just so high above the Horizon of any Place, as is that Place's Latitude.

EXCELLENTLY Explain'd, Madam, said I, and yet I fancy you want to be told further, that the Height, or *Altitude of the*

the Pole Star, as well as *all other Altitudes* of the Sun or Stars, is taken by an Instrument, which hath a Circular Edge like this graduated Meridian, divided on Purpose into *Degrees*, *Minutes*, &c. with Sights fitted to it, to look up at the Object.

I WAS just going to ask you about that, said she; for I remember to have often seen you peering up at the Stars, or catching the Sun-Beams with just such a kind of thing as you describe: But, pray, what Use is this Meridian of, on the Celestial Globe?

THERE, Madam, said I, it shews the *Declination* of the Sun or Stars, by bringing the Sun's or Stars Place in the Ecliptick on the Globe to it, as we did the Places on the Earth upon the other Globe, to find their Latitudes.

Sun's Declination.

Declination! said she, there's a *new Word* for me to learn! which I suppose the Astronomers have coined, to avoid that of *Latitude*; which, when it relates to the Stars or Planets, I think you told me regards the *Ecliptic* only: Well! I doubt my Head will never retain the Memory of all these *Cramp Terms*.

YES, Madam, ſaid I, very eaſily, when you ſo perfectly underſtand their Meaning, for we only forget what we underſtand but by Halves; things thoroughly known become Part of our Nature, as it were; and People can alſo generally remember what they have a mind to. But, however, if you pleaſe to look over Dr. *Harris*'s little Book of the Globes, you may have your Memory refreſh'd at any Time very briefly, and yet plainly and fully.

I THANK you, Sir, ſaid ſhe, for that Information; I ſhall, I hope, be able to underſtand a little of Books of this Kind, by Degrees: But, pray, have you any thing more to ſhew me, relating to theſe Circles?

Greater and Leſſer Circles of the Sphere.

MADAM, ſaid I, 'twill be proper for you to know, that as our Aſtronomers make *ſix greater*, ſo they make alſo *four leſſer* Circles of the Sphere; two of which they call the *Tropicks*, and the other two the *Polar Circles*. The Meaning of the Word *Tropicks* is, *returns back again*; for indeed neither the Sun ſeemingly, nor the Earth really, goes any further in its Annual Courſe, to the Northward or Southward of of the Equinoctial than 23 Degrees and $\frac{1}{2}$; but after it hath gone so far, *returns* again toward it: And because the Points in the Heavens, where these Returns are made, are under the Beginning of the Signs ♋ *Cancer* and ♑ *Capricorn*; they suppose two Circles to be there drawn in the Heavens and on the Earth, parallel to the Equator; and the most Northern of these, and which therefore is our Summer *Tropick*, is called the Tropick of *Cancer*, and the Winter, or Southern one, *that* of *Capricorn*; because they always fall at the Beginning of those Signs.

I LIKE our Earth mightily, said she, for her Steadiness in her Way, and for her not going too far North or South towards the Poles: I love moderate Weather, and would have it be in neither of the Extreams of Heat nor Cold: But, Sir, this Matter now begins to clear up to me apace; when the Sun is in the Northern Tropick, I see our Days are at the longest, and all of them longer than our Nights, during the Time of his whole Stay on the North Side of the Equinoctial: Whereas the very Reverse, I see, must come to pass, while the Sun is on the Southern Side of the Line. But, pray, of what Use

Polar Circles and Tropicks.

Use are the *Polar Circles?* for I see they are drawn on both Globes, as well as the *Tropicks*, and just as far from the Poles as the Tropicks are from the Equinoctial.

OF no very necessary Use, Madam, said I, but only to help to distinguish the Terrestrial Globe into the five Parts, which the Ancients called *Zones*, and which they fancied to be like so many *Girdles* or *Belts* (as the Word *Zone* signifies) encompassing the Earth.

Zones.

O PRAY, said she, let me have some true Knowledge about these *Zones*, for I have heard and read a good deal of them, without being a Jot the wiser.

Torrid.

THE great Space on the Earth, said I, Madam, which lies between the two Tropicks, having the Equator passing thro' the middle of it, the Ancients called the *Torrid*, the *Fiery* or *Roasted Zone*; for they fancied the Sun, keeping always over it, had such a Power here, as to have burnt all things up; and because they had no Knowledge of it, concluded it not inhabitable; whereas 'tis now known to be very comfortably so: Tho' no doubt warm enough to those Inhabitants of it to whom the

the Sun is ſucceſſively vertical, or directly over their Heads, as you eaſily ſee by the Globe he will be.

YES, yes, ſaid ſhe, I underſtand that very well; but I can't help reflecting upon the *Arrogance*, as well as Ignorance, of the Ancients, in ſuppoſing *their* Knowledge to be the Bounds of all things; and glad I am that *we* know ſomething which *they* did not; for I have heard them ſo much cried up, now and then, by Authors, that I could almoſt wiſh my ſelf to have lived among them; but I will, at laſt take Comfort, and thank God that I am a *Modern*, and alive now.— But pray go on about your Zones.

THESE two Spaces of the Earth, ſaid I, Madam, which lie between the Tropicks and the Polar Circles, each Way North and South, the old Geographers called the *Temperate Zones*; and as theſe Oriental Sages, and the Learned *Greeks* and *Romans*, lived (as you (*a*) ſee here) in *one* of them, ſo they did allow the *other* to be habitable alſo.

Temperate.

(*a*) Here on the Terreſtrial Globe I ſhewed her the chief Places of the *Græcian* and *Roman* Empires.

THAT was pretty good-natur'd, ſaid ſhe, for I ſuppoſe they never ſaw the South-

Southern *Temperate Zone*, any more than the *Torrid* one.

NOT that we can find by Hiſtory, ſaid I, Madam: But to proceed; Theſe ſmall Spaces of the Earth, between the Polar Circles and the Poles, they called the *Frigid Zones*, and did pretty juſtly ſuppoſe them not to be habitable, upon the Account of their Coldneſs; for tho' we have ſince diſcovered, that 'tis poſſible to ſubſiſt, and ſeveral of our Ships do yearly go within the Northern Frozen Zone, yet I can't commend it to you as a Place much worth your Enquiry after.

Frigid.

O! don't ſpeak any more about them, ſaid ſhe, you make me ſhiver all over with the Thought of them, and my Blood is juſt going to curdle in my Veins; no *Lapland* or *Spitsburghen*; no Whale-Fiſhing Voyages for me!

YOU ſeem to be really a cold with the Thought of it, Madam, ſaid I; let me warm you a little with this Deſcription of theſe Zones given by Mr. *Dryden*,

From VIRGIL and OVID.

Zones.

Five Girdles bind the Skies: The Torrid Zone
Glows with the paſſing and repaſſing Sun;

Far

Far on the Right and Left the Extreams of Heaven,
To Froſts and Snows, and bitter Blaſts are given;
Betwixt the midſt and theſe the Gods aſſign'd
Two Habitable Seats for human Kind;
And croſs their Limits cut a ſloping Way,
Which the twelve Signs in beauteous Order ſway:
And as five Zones the Ætherial Regions bind,
Five correſpondent are to Earth aſſign'd;
The Sun with Rays directly darting down,
Fires all beneath and frys the middle Zone:
The two beneath the diſtant Poles, complain
Of endleſs Winters and perpetual Rain:
Betwixt the Extreams two happier Climates hold,
The Temper that partakes of Hot and Cold.

WELL, ſaid ſhe, theſe Verſes have a little recovered my Spirits, as well as refreſhed my Memory, and will, I find, fix in the latter, the obliging Pains you have taken to inſtruct me: But pardon me, Good Sir, if I ſtop you a Minute: Mr. *Dryden* here mentions the the Word *Climates*; Pray what are they?

Climates.

MADAM, ſaid I, you will find a deal of uſeleſs Stuff in ſome Introductions to Geography, *&c.* about theſe *Climates*; but all that is neceſſary to know of them, is, that the Ancients ſuppoſing two Circles to be ſo drawn parallel to the Equator, on the Terreſtrial Globe, or at that Diſtance

Parallels.

one

one from another, that to ſuch as inhabit the *leſſer*, the longeſt Day, would be a Quarter of an Hour longer, than it is to thoſe who dwell in the *larger*: Then the Space on the Globe, between theſe two, they called a *Parallel*, and the Double of ſuch a Space a *Climate*; you will eaſily ſee therefore, that theſe Climates muſt leſſen as you go each Way from the Equator to the Poles, and muſt be 24 in Number.

WELL! ſaid ſhe, I ſhall not trouble my Head about reckoning theſe *Climates*; but I think I underſtand what is meant by ſuch a Place lying in ſuch a *Climate*, as well as what the Navigators mean by ſailing in ſuch a Parallel, and that will be enough for me at preſent; but I will tire you no longer now, I'll get the Book you adviſe me to, which I believe I have above among my Brother's things; and after I have conned my Leſſon well over, you muſt expect that I ſhall ask you abundance of Queſtions more.

WITHIN a ſhort Time after this, the Ingenious and Inquiſitive Lady got her Globes ſet out again, and began with me thus:

I HAVE been looking over the little Book you recommended to me, Sir, ſaid ſhe, which I think is very plain and conciſe, and I fancy I am now got to be ſuch a Proficient, as that I am qualified to go thro' the *Problems*, as the Book calls them, tho' what that Word ſignifies I don't underſtand.

Problems.

THAT *Greek* Word, Madam, ſaid I, ſignifies *ſomething to be done or practiſed*, and I queſtion not but you have ſo well conſidered this Affair, as to be able to work or perform any of theſe *Problems* upon the Globes your ſelf.

I DON'T know that, ſaid ſhe, but I'm reſolv'd to try, and with a little of your Help, perhaps, I may get thro' them: Come, pray, let's begin; and, *firſt*, ſhew me how to *rectify* each Globe, as he calls it, and what I ſhall learn by that.

Rectifying the Globe.

Rectifying the Globes, Madam, ſaid I, is reducing them to ſuch a Poſition, as that they ſhall truly repreſent the Situation of the Circles of the Sphere of the fixed Stars and Planets; and of the Poſition of the Earth itſelf at any Time aſſigned.

VERY

VERY well, said the Lady, let us then take his Time of the Year; suppose *May* 10, 1719; How must we begin?

MADAM, said I, for common Use, look first for the Sun's Place, against the Day of the Month, in the Calendar, on the wooden Horizon (tho' if you would proceed to greater Exactness, you must find the *Sun*'s *Place* in some good Tables, such as those which Dr. *Harris* hath given in the second Volume of his *Lexicon*, or such as *Parker*'s Almanack which I have here in my Pocket, gives you every Year, or else you must determine it by Calculation, *&c.*) and then finding that Place, or what Degree of any Sign of the *Zodiac* the Sun appears to be in that Day at Noon, which you will find to be then in in the first Degree of *Gemini*, look it out on the *Ecliptic* on the Globe, and there make, either with a Pencil or with Ink, a Mark to represent the Sun for that Day.

Sun's Place.

BUT, said she, won't that spoil the Globe?

NO, Madam, said I, *that* being varnish'd, the Ink will easily come out again, if you rub it with your Handkerchief a little

little wetted; as ſoon as this is done, you may alſo, if you pleaſe, by the Help of *Parker*'s or ſome ſuch *Ephemeris* or Aſtronomical Diary, place *all the Planets* on your Globe, after the ſame Manner, allowing for their Latitude, either North or South, of the Ecliptic.

Thus the Moon being then in 24° 33′ of *Cancer* ♋, and having about 4° 41′ of South-Latitude, take, with a Pair of Compaſſes, thoſe Degrees and Minutes of Latitude from the Meridian, or any great Circle, and placing one Foot in 24° 33′ of ♋, turn the other directly towards the Equinoctial, and there make this Mark ☽ to repreſent the Moon.

After the ſame Method you may place ♄ *Saturn* in 8° 42′ of *Virgo* ♍; and ♃ *Jupiter* in 26° 35′ of *Leo* ♌: Then make alſo this Mark ♂ for *Mars* in 16° 32′ of *Aquarius* ♒: And this Character ♀ for *Venus* in 28° 18′ of *Gemini* ♊: Laſtly, placing *Mercury* ☿ in 11° 4′ of the ſame Sign, you will have adorn'd your Globe with the Characters of the Seven Planets, all appearing in their proper Place as they are in the Heavens.

THIS is mighty Entertaining, ſaid ſhe; hre take this Pencil quickly, and let me ſee you juſt now place all your Planets upon the Globe according as they ought to be done, that I may learn how to range them another Time: For I fancy their very *Characters* or Figures ſo much, that I could almoſt wiſh our Patches were cut into ſuch pretty Forms; but that I fear 'twill revive the fooliſh Notions of Aſtrology again, which you have taught me to deſpiſe. But, pray, continued ſhe, how do you know the Planets from the fixed Stars when you ſee them in the Sky?

PRETTY eaſily, ſaid I, Madam, as to *Saturn*, *Jupiter*, *Mars*, and *Venus*. And *Mercury* is ſo near the Sun as to be very rarely ſeen at all.

THAT puts me in Mind, ſaid ſhe, of what Sir *Richard Blackmore* ſaith of him in his Poem called *Creation*, in theſe Lines.

> Mercury, *neareſt to the Central Sun,*
> *Doth in his oval Orbit circling run;*
> *But rarely is the Object of our Sight*
> *In Solar Glory ſunk, and more prevailing Light.*

WELL remember'd, Madam, ſaid I, But to our preſent Point, the Knowledge of theſe Planets from the fixed Stars: The former, you muſt know, don't *twinkle* as the

the fixed Stars do; beſides they are always and all of them in or near this Line here called the *Ecliptic*: Which you may eaſily learn to trace out in the Heavens, by theſe Conſtellations which compoſe the Twelve Signs; and if you ſhould, at laſt, doubt about the *Planets*, if you ſee them change as they will do, in ſome Time, their Diſtance from any fix'd Star that you know; you may eaſily diſtingiuiſh them to be *Erraticks* or Planets.

I THINK, ſaid ſhe, you reckon'd ſeven Planets juſt now; ſure I have read ſome where, that there are more.

IN that Account above, ſaid I, Madam, I followed only the Vulgar Way of Computation, for in Reality the Sun is no Planet or Wanderer, but a fixed Star placed in the Center of our Syſtem, and in all Probability like the reſt of thoſe that we ſee in the Heavens. And round him, as a Center, *Mercury*, *Venus*, *Mars*, the *Earth*, *Jupiter*, and *Saturn*, do revolve, and are now called *Primary* Planets; becauſe they revolve round the Sun, as their Center: While the others we call *Secondary Ones* or *Satellites*, i. e. Guards or Attendants, becauſe they revolve round ſome one of the *Primary* Planets,

Number of Planets.

nets, as their Center, and together with it, move alſo round the Sun.

Thus the Moon is a Secondary Planet, whoſe Center of Motion is our Earth, on which ſhe conſtantly attends, and her Circle round us ſhe performs in about a Month's Time, while at the ſame time, ſhe revolves together with the Earth round the Sun in its Annual Courſe. *Jupiter* hath four ſuch Moons or *Satellites*; and *Saturn* five, revolving round Him: But it doth not yet appear that *Venus* or *Mars* have any *Satellites* at all.

Satellites.

As for *Mars*, ſaid the Lady, I ſhan't trouble my Head about him; tho' one would think, the God of War, or Captain-General of Heaven, might command a few Guards or Followers: But I will never forgive the Aſtronomers, nor believe at all in Teleſcopes, if they don't find out that *Venus* hath ſome *Attendants*; that is ſuch an Affront to our Sex, as we muſt never paſs by. But to be ſerious, I ſuppoſe, *Mercury* and *Venus* being ſo near the Sun, have no occaſion to be lighted in the Night by Moons, as the more remote Planets have; tho' why our Earth ſhould have one, and yet *Mars* none, is not, methinks, ſo eaſy to be accounted

counted for. But we have made a long Excurſion from our Globes; pray let's return to them: And let me ſee what I ſhall be the better for knowing how to *rectify* the Globes, and to *patch* on the Planets, as you juſt now have ſhewed me the Way of.

Hour Circle and Index.

MADAM, ſaid I, bring the *Sun's Place*, for *May* 10, to the graduated Side of the Meridian, and then turn or ſet the *Index* of the Hour-Circle (placed here as you ſee upon the Braſs *Meridian* about the Pole) to Twelve at Noon; and then your Globe will be fitted to ſhew you the State of the Heavens. As it now ſtands, the Mark for the Sun repreſents his being on the Meridian, as he is every Day at Noon; and there it will ſhew the Sun's *Meridian Altitude* above the South Part of the Horizon to be 58 *Degr.* 42 *Min.* Then if you will bring that Mark to the Eaſtern Edge of the Wooden Horizon, you will ſee there *what Point of the Compaſs* the Sun riſes upon, and your *Index* will ſhew you the Time of it; and if you bring the Sun's Place to the Weſtern Edge, you will find how far from the true Weſt Point the Sun ſets, and what a Clock it is when he *goes down*, as we call it: Thus, *May* 10, the Sun riſes about $\frac{1}{4}$ of an Hour before 5 a Clock; and sets $\frac{3}{4}$ of an Hour after 7.

WELL, said she, I fancy I shall be able to make an Almanack in a little Time.

THAT you may soon do, this Way, said I, Madam, and much better than most of those who publish them: But if you have a Mind to know the Stars and Planets, how they will appear, and are situated at any particular Time, suppose to Night at Eleven a Clock; you need only turn the Globe about till your *Hour-Index* points to Eleven at Night; and then putting a little Piece of Paper under the Brass Meridian, to stay the Globe in that Position, please to turn the Frame, and Globe and all, about, till the North-Pole here point up towards the *Pole Star* in the Heavens; and then you will have all you can wish for shewed you; for, by comparing the Pictures and Marks of the Stars and Planets with the real Ones, at that Time in the Heavens, you will find them exactly to answer to one another; and *these* on the Globe will make *those* easily and sufficiently known to you.

SIR, said she, after abundance of Thanks, I must beg you to break off

When

here; we mnſt defer this till Night:

When with the Stars we'll be familiar,
As e'er was Almanack Well-willer.

And in the mean time, I'll con my Leſſon in the Book, that my Ignorance may not give you too much Trouble. The Tea waits us; will you pleaſe to move, Sir?

THE Evening of this Day proved one of the fineſt I ever ſaw, and the Night ſucceeding it was ſo very clear and bright, that the Moon being then not above our Horizon, there appeared many more Stars than uſual. As we were walking to a Summer-Houſe, placed on a Mount in the Garden, where the Lady had order' the Celeſtial Globe to be ſet out, ſeveral Poetical Deſciptions of ſuch a Night occurred to our Thoughts, and were recited. The Lady cloſed all with that famous one of Mr. *Dryden*,

All things are huſht, as Nature's ſelf lay dead,
The Mountains ſeem to nod their drowzy Head,
The little Birds in Dreams their Songs repeat,
And ſleeping Flowers beneath the Night-Dew
Even Luſt and Envy ſleep,——   *(ſweat:*

I was going to ſay —— *But Love denies*, &c.

 when

when ſhe interrupted me, and ſaid, I'll have nothing of *Love* mention'd nor talk'd of to Night; the Opportunity is too ſolemn, and I'm affraid I ſhall grow in earneſt and ſerious about it: We will both make our Court now only to *Urania*, and every *gay thing* ſhall give place to *Aſtronomy*: Let's enter the Summer-Houſe, and ſee whether I have rectified the Globe as it ſhould be, and ſet it right to repreſent the preſent Time, which is juſt half an Hour paſt Ten.

MADAM, ſaid I, you have done it with Accuracy: And I ſee you have mended the haſty clumſey Figures, that I had made, of the Planets, and have placed very beautiful ones, of your own, in their Room.

BUT, ſaid ſhe, I don't know how to place the Globe due North and South, as my Book directs, unleſs there were a little Compaſs here, placed on the Frame.

MADAM, ſaid I, there uſually is ſuch a Compaſs made on purpoſe to be placed on the Globe; but I can ſhew you how to ſet the Globe right enough without it; you ſee theſe 7 large Stars here, that are painted within the Figure of the greater Bear,

Charles-wain.

3 in the Tail, and 4 in his Body: Theſe our *Engliſh* Country People call *Charles-Wain*, and fancy the *four* to be the 4 Wheels of the Waggon, while, forſooth, the *three* are to repreſent the 3 Horſes that draw it. But as to the preſent Concern, pleaſe to take Notice, that as this Conſtellation, in our Horizon, *never ſets*, but ſeems to revolve round the Pole in 24 Hours; ſo theſe two Stars of the 7, that are neareſt to the Pole Star, or the two hinder Wheels of the Wain, do always *point* up pretty nearly to the Pole Star; and are therefore ſometimes called the *Pointers*; and conſequently, if you carry your Eye on in a Right Line from them, they will direct you to the *Pole Star*, which you ſee is here, on the Globe, placed in the End of the Tail of the *Leſſer Bear*, a Conſtellation of 7 pretty large Stars, much in the ſame Figure of thoſe in the *Great Bear*, or *Charles-Wain*.

Leſſer Bear and Pole Star.

I SEE them on the Globe, ſaid ſhe, let us now look out of the Window and obſerve them in the Heaven; O! I ſee them yonder very plain, ſaid ſhe, and now I ſhall know in the Night as well as the Day, how to find the four Points of the Compaſs, *Eaſt*, *Weſt*, *North*, and *South*.

WE muſt then return again to the Globe, Madam, ſaid I, and by opening the North Window, direct its Pole to point up to the *Pole Star*, and ſo ſet it as near as we can due (*a*) North and South.

(*a*) Here the Braſs Meridian of the Globe was placed due *North* and *South*.

There is no need of great Accuracy for our preſent Purpoſe; and I think it ſtands pretty true now. Before we look or go out again, pray, Madam, pleaſe to obſerve this Situation of the Globe, and then you will eaſily ſee how the Poſition of the Stars do at preſent correſpond with it: There is indeed, now not any very eminent Star, or one of the firſt Light or Magnitude, *exactly* on the Meridian, either North or South: But you will ſee this great Star, which is called the *Virgin*'s *Spike*, becauſe painted on an Ear of Corn which ſhe holds in her Hand, a little to the Weſtward of the South, and about 28 Degrees high above the Horizon; as you ſee, appears by bringing this Quadrant of Altitude, ſcrewed in the Zenith, to it; whikh is an Arch of 90°, and being moveable, ſerves to ſhew the Altitude of any Star or Planet.

Spica Virginis.

I SEE that, ſaid the Lady, *here* on the Globe; But how ſhall I be able to find and count Degrees in the Heaven?

YOU

YOU know, Madam, said I, that it hath been before observed to you, that the Astronomers have Instruments made on purpose for it, which do it with great Accuracy: But as for your present Enquiry, how high any Star or Planet appears to be above the Horizon, you may guess at it nearly, thus: The Distance you see here between the *two Pointers* of the *Great Bear* before-mentioned, is nearly five Degrees; and this being a Distance always ready, and in view, will serve you very well to guess at the Height of any Star above the Horizon; or at the Distance of one of them from another; so as to enable you to find out any of them in the Heavens by the Help the Globe, or any Planisphere, or Map of the Heavens: Use will make this easy to you; and when you come also to consider, that from the Zenith to the Horizon, being 90°, half that Distance mnst be 45°; one third of it 30°; a sixth of it 15°; a ninth Part of it 10°, *&c.* you will, by Degrees, easily gain a practical Knowledge of these Distances.

But if you please we will go on: Almost South-west, at this Time, and about 43° high, will appear another Star of the first Magnitude, called *Deneb*, which is in the

Deneb.

the Tip of the Tail of the Lyon; I ſee it yonder ſimpering thro' that Weſtern Window; if you will let me lift up the Saſh you may ſee it without going out.

O! I do, ſaid ſhe, and the Virgin's Ear of Corn too, very plain: But what are thoſe two great Stars that appear together almoſt nearly Weſt, and let me ſee!—don't tell me — about, about—-I muſt look out at the *Pointers* again to get my Meaſure —— why, they are about 25 Degrees high.

VERY well gueſs'd, ſaid I, Madam; you will come to meaſure the Diſtance of Stars by your Eye, in a little Time, as accurately as the good Houſwives and Workwomen can meaſure Cloth or Ribbons, by the length of their middle Finger.

WELL, ſaid ſhe, Mr. *Obſervator*, and ſo I can too, for all I have a Mind to be an Aſtronomer, as well as the beſt of them; and I don't deſign, Sir, that my Studies ſhall ſpoil my Houſewifry: But pray tell me quickly, who thoſe two famous Stars are.

THE

THE uppermost, Madam, said I, is called the *Lyon's Heart*, and is you see drawn here on the Globe: And the other is *Jupiter*; you remember you have drawn the Character of him here your self.

Cor Leonis.

BLESS me! said she, is that *Jupiter*-- well, I have many Questions to ask you about that Planet another Time, but I will not stop you now; pray go on, and shew me how to know more of these Stars and Planets; for I begin to grow mighty fond of their Acquaintance.

DON'T you see, Madam, said I, here on the Globe, two Stars, about 15 or 16 Degrees high, and within two Points to the Westward of the Northern Edge of the Horizon: These two are called the Shoulders of *Auriga*, and the lowermost and most Northern is called *Capella*, and is a Star of the first Magnitude; these are very conspicuous Stars, and you may see them in the Heavens very plain out of that Northern Window.

Capella.

I DO, said the Lady, very clearly, and I see, said she, also another pretty remarkable Star, about the same Height with

with *Capella*, about a Point to the Northward of the West, under the *Great Bear*; pray, what Name do you give him.

THAT is called *Pollux*, Madam, said I, and his Brother *Castor*, you see, sits here close by him on the Globe, and between them they make up one of the Signs of the *Zodiac*, which they call *Gemini*, or the Twins.

Is that, said she, the Deity that the Countrywomen swear by, when they cry O, *Gemini!* ——But don't look grave, or give me any Return: For tho' I trifle, and am Impertinent, I won't allow you to be so. Let us go on and see what noble Stars we can find to the Eastward of our Meridian.

MADAM, said I, if you will look out at that North Window, and direct your Eye along by the Pointers of the *Great Bear*, till you see past, or beyond, the *Pole Star*, and continue it down till you come within 20 or 30 Degrees of the North North-East Part of the Horizon, you will see an Eminent Constellation which is called *Cassiopæia*'s *Chair*: This is the Figure of it here on the Globe; 'tis always

always opposite to the greater Bear, either above or below the *Pole Star.*

I SEE it, said she, very plain, and a very notable Collection of Stars it is; but, pray, said she, what do you mean by calling it *Cassiopæia's Chair*, who; or what was that *Cassiopiæa?* sure I have read something about her, in some Books of the Heathen-Gods.

Cassiopæia.

NO doubt of it, Madam, said I, and the Company you will see she is in, will refresh yourMemory. This *Cassiopæia*, the Poets tell you, was the Wife of *Cepheus*, who was, once upon a Time, King of *Æthiopia*; and *here* the good old Monarch stands upon the Globe, with his Scepter in his Hand, just above *Cassiopæia*; and below her, at the very Edge of the Horizon, you see, you are to look for her fair Daughter *Andromeda*, who had the Vanity to think herself handsomer than the *Neriedes* or Sea-Nymphs, which put them into such a Rage, that they immediately applied to old *Neptune*, the God of the Sea, to revenge the Indignity.

Cepheus.

Andromeda.

On this, the obsequious Deity sent a huge ugly Monster up into the Country, which did great Mischief there: The poor People, who in those Days were always

punished

puniſhed for the Sins of the *Great Ones*, apply'd to the Oracle for Relief, and were told that the only way to appeaſe the Gods, who were all on the Side of the *Nereides*, was, to expoſe the audacious *Andromeda* to be devoured by a Sea-Monſter; which I ſuppoſe *Neptune* undertook to get ready for that Purpoſe: This was done, but the gallant *Perſeus*, whom you ſee here on the Globe, juſt behind her, as her Champion, deliver'd her and kill'd the Monſter, and I hope carried off the Lady; and to reward the Mother of ſo beautiful a Creature, he got *Jupiter* to *ſtick her* up here among the Stars, and they form the Celeſtial Chair in which ſhe ſits in State: And, thus, Madam, I have given you the Hiſtory of one of the Conſtellations, and if you pleaſe, I can tell you as *long* and as *true* a Story of many of the reſt.

Perſeus.

I THANK you, Sir, ſaid ſhe, but you ſhall not, this is enough for a Sample; and now I remember all this Stuff about *Perſeus* and *Andromeda*, as well as if I had ſeen the whole Affair, as I believe I did once, or at leaſt good Part of it ſhewn upon the Stage: And have much oftener ſeen it in Pictures and Prints. But dropping

ping all Fables, let me go with my Lesson, I shall know *Cassiopæia* again, whenever I see her.

BUT, said I, Madam, I have a Story to tell you of *Cassiopæia*'s Chair, that is no Fable, but a certain Truth, and yet is equally strange with the other fabulous Relation. About the Year 1572, there appeared a *New Star* in this Constellation, which at first was as big as *Jupiter* appears now to be, and was fix'd to one Place like the rest of the fixed Stars; but lessen'd by Degrees, and at last, at the End of 18 Months, went quite out, and appear'd no more.

THAT indeed is a very unaccountable thing, said the Lady, but as I have met with some such Relations of other fix'd Stars, so I shall leave my Surprize, and my Queries about it, till I come to trouble you about the Nature, Uses, and Distances of the fixed Stars in general; for I must have a deal of Talk with you about *that* and other things in Astronomy, before you get quite rid of me, and you must thank your self, if my Curiosity be teazing and impertinent, for you have wound it up to a very great Height I'll assure you. But, pray, Sir, let us now go on and make an end of our Stars, it grows late and the Air cold.

MADAM, ſaid I, we ſhall diſpatch the reſt, as faſt as you pleaſe, for the Way I have ſhewn you, of finding and diſtinguiſhing the Stars above-mentioned, will teach you to do ſo, by any others in the Heavens: Thus you will ſee here above the Pole-Star, and about 14 Degrees from him, and a little to the Eaſtward of the Meridian, the Conſtellation, called the *Little Bear*, conſiſting alſo of ſeven pretty eminent Stars, of which the lowermoſt, *now*, or that in the Tip of his Tail, is the *Pole-Star*: You ſee here, almoſt due Eaſt, a fine bright Star of the firſt Magnitude, which is called *Lucida Lyra*; and under it, a little to the South of the Eaſt, as *Lyra* is to the North, another great Star of the firſt Light, about 12 Degrees high, which is called *Alcair*: And you can't but take Notice of theſe four Stars here all of the ſecond Magnitude, placed in the Form of a Lozenge, which is called the *Dolphin*. About 8 Degrees high, and about 2 Points and ½ to the Eaſtward of the South, you ſee alſo a famous Star, of the firſt Light, in the Body of *Scorpio*, one of the 12 Signs.

Urſa Minor.

Lucida Lyra.

Alcair.

Dolphin.

Scorpion's Heart.

ALL

All theſe Stars I ſee, ſaid the Lady, and I think diſtinguiſh very well; and I fancy I ſhall be able, by Degrees, by the Help of ſuch eminent Stars as *that*, which I ſee here on the Globe, are placed pretty near the *Ecliptick*, to *trace out*, as you ſaid a while ago, that Circle in the Heavens. But, pray, firſt tell me, what you call that Star, or rather Planet, (for I fancy 'tis one of thoſe *wandring* Lights) which appears yonder, almoſt upon the South Part of the *Meridian*, and about 25 Degrees high.

You have gueſſed very right, Madam, ſaid I, 'tis a Planet, and the moſt remote one of all, *Saturn*.

Saturn.

Is that *Saturn*, ſaid ſhe, I'm heartily glad to ſee him, I ſhall know him again another Time; I long to peep at him thro' a Teleſcope, and to ſee his famous *Ring*. But of this, more ſome other Time, when the Teleſcope, you have promiſed me, is fitted up: Is there any thing elſe worth obſerving, before we remove to our Sleep?

Only pleaſe, Madam, ſaid I, to take Notice of that *Track of Light*, yonder in

Heavens, and here drawn upon the Globe, which is called the *Milky Way*. You see here by the Globe, as the Position of the Heavens is now, that it begins at the North Part of the Horizon, about *Perseus*, takes in *Cassiopaeia*, and after that the *Swan*, and then runs on toward *Scorpio*, and towards the *South Pole*, and takes in the famous Constellation called the *Cross*; then it turns *Northward* again, thro' the *Ship*, a little above the *Great Dog-Star*, or *Syrius*, and above the Right Shoulder of *Orion*, and thence taking in *Capella*, runs on towards *Perseus*, where we began to trace it.

Milky Way.

I'M glad you thought to shew me this, said she, before we finish our Night's Observations; I see it plain in the Sky, and perceive that its Figure, on the Globe, corresponds exactly with it; I won't stay now to ask you what it is, because that may be one of my many Questions to you another Time; we will only remember what Mr. *Milton* saith of it:

A broad and ample Road! whose Dust is Gold,
And Pavement Stars, as Stars to us appear,
Seen in the Galaxy, that Milky Way,
Like to a circling Zone powder'd with Stars.

Mr.

Mr. *Milton*, Madam, ſaid I, alludes to the Notion that the Poets had of it; that it was the Path which the Gods uſed in the Heavens, which Mr. *Dryden*, from *Ovid* thus alſo deſcribes:

A Way there is in Heaven's extended Plain,
Which when the Skies are clear is ſeen below,
And Mortals, by the Name of Milky, *know:*
The Groundwork is of Stars; thro' which the Road,
Lyes open to great Jupiter's *Abode.*

SIR, ſaid the Lady, a Thouſand Thanks to you for the Pains you have taken to inſtruct me, and I wiſh you a good Repoſe.

THE next day my Affairs called me away, for ſome Time; but at my Return, as I found the Teleſcopes and other Inſtruments, I had ſent for, in perfect good Order, ſo I found the Lady had been cloſe at her Aſtronomical Studies: She was exactly ready in all Problems upon either Globe, and had gotten ſuch an intimate Knowledge of the Stars, that ſhe had alſo acquired a very tolerable Knowledge in the ſeveral *Syſtems* of the Univerſe, or *Hypotheſes* to ſolve the Celeſtial Appearances, as they are called by

Aſtronomers; and long'd with great Impatience to ſee the Uſe of the *Teleſcopes* and *Quadrants*, &c. which I had ſent down to her Country-Seat.

Where ſoon after I arrived, ſhe put me upon beginning our Obſervations, and had methodized the Enquiries and Queſtions ſhe deſigned to make, with great Addreſs and Dexterity.

Sun.

LET us begin, ſaid ſhe, to talk a little about the *Sun:* I think you agree, that his Centre appears to move always in the ſame Line, or in the *Ecliptick*; but I think you ſay his apparent Motion is *unequal* there.

'TIS true, ſaid I, Madam, for when the Earth is *neareſt* the Sun, as ſhe is in Winter, then ſhe, in *reality*, and the Sun, *ſeemingly*, moves *faſter* than in the Summer, when the Diſtance between the Earth and the Sun is greater; and accordingly the Sun's Diameter appears *bigger* in Winter than in Summer.

BUT, Sir, 'tis ſtrange, methinks, ſaid ſhe, that the Sun's *nearer Approach* to us in Winter than in Summer, doth not counterchange thoſe Seaſons: Have not we we the greateſt Heat from the Sun when we are neareſt to him?

No, Madam, ſaid I, for the different Heat of our Seaſons of the Year, do not depend upon *that*, but upon the Sun's Rays falling more *directly*, or more *obliquely* upon us; for in the Diſtance of 70,000,000 of Miles, a little Approach of the Earth to, or its Receſs from the Sun, will make no ſenſible Alteration as to Heat or Cold. But there is another thing ariſing from this Inequality in the Earth's Motion round the Sun, which is pretty conſiderable, and that is, that the Sun will appear to tarry about 8 Days longer in the Northern Part of the Ecliptick than he doth in the Southern; the Reaſon of which is, from the Figure of the Earth's oval or elliptick Orbit: [*See* Lexic. Techn. *Vol.* II. *under* Sun.] And thus having given you ſome general Ideas of theſe things, I wait your further Commands.

WHY, then, ſaid the Lady, pray give me now, for it ſeems to be a proper Place, ſome little Knowledge about the *Equation of Time*, which I have read a good deal about, and Tables of which I have ſeen hanging by Clocks, and put upon Dials

 and

and Watches; Pray, can our Sun be in the wrong, don't he measure Time equally?

MADAM, said I, the daily Revolutions of the *Earth*'s Equator round its Axis, are exactly equal in Time to one another; and yet the Time from the apparent Noon of one Day to that of the next, is unequal, and sometimes greater and sometimes lesser.

WELL, said she, I'm glad however, 'tis not our Earth's Fault, and that she is so regular in her diurnal Whirls: But, pray, let me then know, where the Error, or Inequality, lies?

THERE is, said I, Madam, a double Cause of this Inequality; the former is, that the Earth's Annual Orbit is not an *Exact Circle*; and the other is, that the Earth's *Equator*, about which the Diurnal Motion is made, and the *Ecliptick*, or the Circle she describes round the Sun, are not *co-incident*, or in one and the *same Plane*, but make an Angle, as you know they do, at their Intersection, of 23° 30′ of which when your Curiosity, and further Knowledge of these Affairs, leads you to make more full Enquiries, you will

will receive a plain and ſatisfactory Account, from Mr. *Whiſton*'s Aſtronomy, p. 116, 117, *&c.*

I THANK you, Sir, ſaid ſhe, but, pray, let us now get all things ready to look on the Sun, with your Glaſſes, that I may know, by my Eyes, as much as I can of that wonderful Luminary, the great Centre of all the Planets Motions.

OUR Teleſcope was about 14 Foot long, and had a plain Glaſs, ſmoaked with a Candle, ſcrew'd on before the Eye-Glaſs, to defend the Eye from receiving any Injury from the too intenſe Light of the Sun.

AFTER ſhe had look'd upon him 2 or 3 times; It appears plainly, ſaid ſhe, to be a great Globe of Fire, or rather, as *Butler* ſaith,

———————— A Piece
Of red hot Iron, as big as Greece.

and ſo no doubt it muſt be, by the great and conſtant Heat which it gives: But, pray, tell me, as fully as you can, what the late Aſtronomers and Philoſophers have

have discover'd about this vast World of Light. I perceive you suppose him fixt and immoveable, as to Place, in the Centre of what you call the *Solar System*; but doth he turn round his own Axis or not? how much bigger is he than our Earth? how far is he from us? and how can his Heat continue so long as it hath done, without any sensible Wast or Diminution?

Solar Spots.

MADAM, said I, by observing carefully the Spots, which often appear in the Sun's Face, tho' there happen to be none now, they have discovered, that the Sun revolves round its own Axis, in about 25 Days.

SPOTS! said she, What, are there *Spots* in the Sun, which sometimes appear there, and sometimes not; for God's sake what are those *Spots*?

THERE are various Opinions about them, Madam, said I, but the most probable one, is, that they are a kind of Dross or Scum which sometimes gathers upon his Face, as is the Case of melted Metals; for I have seen several *Spots*, which for a Time appeared distinct, at last some of them quite vanish'd, and others

others run together into one, and ſo compoſed a much greater Spot, as was the Caſe at the Time of the laſt famous Eclipſe of the Sun; and ſome of theſe Spots muſt be immenſely large, to appear ſo big as they do, ſometimes, to us, conſidering the prodigious Diſtance of the Sun, which probably amounts to about 70 or 80 Millions of Miles. *Diſtance of the Sun.*

EIGHTY Millions of Miles! ſaid ſhe, Why you fright me, my Head turns round, and I'm giddy with the very Notion of it!

AND yet, Madam, ſaid I, as great as this Diſtance is, a Ray of Light runs it in about 7 or 8 Minutes Time; while ſuch a ſwift Traveller, as a Cannon-Ball, ſuppoſing it to move all the Way as faſt as when it juſt parts from the Gun, can't arrive there in 25 Years. Theſe things muſt needs appear wonderful and ſurpriſing to you, but we have *very good Reaſons* to conclude that they are very near to Truth; *which* I forbear to mention, becauſe perhaps at preſent, you may not be qualified fully to comprehend them.

I DOUBT indeed, I am not, said she, which I heartily lament, and I envy you Men and Scholars, as much as I dare, the Pleasure of knowing the Reasons of, and inquiring into the Natures of these amazing things. But, pray, Sir, is not the Bigness of the Sun answerable to this vast Distance that he is from us?

YES, Madam, said I, according to these Ways of Computation, the Sun's Diameter, or his Breadth from one Side to the other, is about 800,000 Miles, which is above 100,000 times greater than the Diameter of our Earth; and therefore his Bulk, or rather the Quantity of Matter in the Sun, must exceed that of the Earth above 10,000,000 times.

And this Consideration of the Vastness of the Sun's Magnitude will account for the Query you rightly enough started, how he can so long continue his Heat without any sensible Diminution? For we take the Sun and the fix'd Stars to be only very great Bodies of Earth, vehemently hot, whose Heat is preserved by their Greatness, and by the mutual Action and Re-action between them and the

Sir *Isaac Newton's* Opticks, p.318. last Edition.

the Light which they emit, and whoſe Parts are kept from burning out and fuming away, not only by their being of a fix'd Nature; but alſo by the Weight and Thickneſs of the *Atmoſpheres* which are round about them, and which do ſtrongly compreſs them, and condenſe the Vapours and Exhalations which would otherwiſe fly away from them; but are now by this Means made to fall back again upon his Body; and as to the daily Expence of his Light and Heat, the Particles of Light are ſo infinitely ſmall, that out of a Body ſo big as the Sun, they may be ſent for many hundred thouſands of Years together, without any ſenſible leſſening of his Bulk, Weight, or Magnitude.

I BELIEVE, I comprehend the Main of your Reaſoning, ſaid ſhe, but I am got a little out of my Depth; let me recover firm Ground again, and then I would aſk you farther, whether, ſince you take the Sun to be an immenſe Globe of Earth, thus ſet on Fire, and the fix'd Stars to be Suns, or Bodies of the ſame Nature, you don't think the Stars, ſeverally, to have the ſame Uſe, and to be the *Centres* each of them, of *Syſtems of Planets* revolving round them, as ours do round

Fix'd Stars *ſo many* Suns.

the Sun, to whom they afford ſuch all-chearing Light, and enlivening Heat, as our Sun doth to us? For methinks 'tis a mean Uſe of them, and below the Wiſdom of our *Great Creator*, to place them in the Sky only to twinkle and divert us; whereas, all of them put together, don't afford us the 10th Part of the Light of the Moon; but if we ſuppoſe them all to be Suns to ſome *other Syſtems* of Planets, becauſe of their vaſt Magnitudes, and becauſe of their ſhining, as I think you agree they do, by their own Light, and not with one borrowed, like that of the Planets; what a glorious Idea doth it give us of the Almighty Power! of the Wiſdom and Goodneſs of the Divine Nature? And what a poor contemptible Opinion ought it to make us entertain of our ſelves, who perhaps may bear as little a Proportion in *Wiſdom* and *Knowledge*, to ſome of the Beings that inhabit the Starry Regions, as we do in *Magnitude* to them all; for I can eaſily conceive infinite Degrees of Knowledge and Perfection, with as great a Variety, that may be in a *Series* in Creatures between us and the Deity; and perhaps there may be alſo a conſiderable one below us.

I AGREE intirely with you in that Speculation, Madam, ſaid I, but we muſt touch it tenderly, or elſe the old Divines will be angry.

AY, ſaid ſhe, ſuch of them as imagine all things made for the ſake of *Man* only; but I have no ſuch lofty Notions of the Dignity of our Species; and I think Mr. *Oldham*'s is a very juſt Satyr upon that narrow Notion, when, with regard to the very Point before us, he ſaith, that Man believes,

That Turnſpit Angels tread the Spheres (for him.

But now I'm talking about the Stars, pray tell me once for all, have theſe Stars and the Planets no real Influence upon us Mortals?

NOT at all, Madam, ſaid I, ſo as by any Phyſical Power to Influence our Wills, Lives, and Actions; that kind of Effect is certainly more true with regard to what the Stars have often been *unequally compared*, the *Eyes* of a fine Lady of good Senſe and Virtue, for *Thoſe* do certainly, like the *Eyes of a good Magiſtrate,*

gistrate, *scatter* and disperse all Evil before them: They heighten our Genius, and inspire us with Wit, and yet keep our Conversation as chast and modest as they make it entertaining and instructive.

I ALWAYS take your Compliments for *Instructions*, said she, and have no Excuse to make for the Vanity of being pleased with them, but that I will endeavour to be as good as you represent us, and we shall have a much better Reason than ever to value the Power we may have over your Sex, if we can make it help us to reform it: But you will not allow me then to believe any Astrologic Influences?

NOT any at all, said I, Madam, for they serve only to nurse Superstition, to fill us with false Fears, deceive us with vain Hopes, and to excite a dangerous Curiosity, and an unreasonable Inquisitiveness into Futurities; and it is indeed, in effect, either making the Stars so many Deities, and consequently running into some of the worst Sorts of the Idolatry of the Heathens; or else 'tis introducing the Notions of a Physical Fatality, and banishing out of our Minds all Religious and Moral Notions.

SIR,

SIR, said she, I acquiesce; and to tell you the Truth, I never had much Faith in things of that Nature: But let us leave this Sun of ours for the present,

This Sun, of our poor World both Eye and Soul
This Sun, that with surpassing Glory crown'd,
Looks from his sole Dominion like a God;
That by magnetick Beams thus gently warms
The Universe, and to each inward Part,
With gentle Penetration, tho' unseen,
Shoots genial Virtue even to the Deep:

As I think *Milton* expresses it; and give me leave to ask you a few Questions about his *Brethren, the Stars.* If the Sun and they be nearly of the *same Bigness*, as they appear to be of the *same Nature*, what an infinite Distance must they be from us?

Fix'd Stars.

'TIS very justly observed, Madam, said I; for indeed, whatever their Bigness be (and much less than the Sun we have no reason to suppose any one of them to be) their *Distance* is so great, that the Diameter of the Earth's yearly Orbit or Circle round the Sun, which you know must be double to his Distance from us, and therefore about 160 Millions of Miles: This I say, according to all the

Their immense Distance.

Observations we can make, and the Reasonings we can form, bears no manner of Proportion, and is but a Point, in comparison of the Distance of the *nearest fixed Star*; for we have no Reason to suppose them *all equally remote* from us. And could we advance towards these Stars 99 Parts in a 100 of the whole Distance, and that there were but one hundredth Part of the present Distance remaining, they would appear very little larger to us than they do now. The Distance of *Syrius* or the *Dog-Star*, Mr. *Huygens* takes to be about 27000 times as far from us as the Sun is; so that I believe we are not much out of our Computation, when we conclude, that a Ray of Light cannot come from thence in less than 6 Months time, nor the Cannon-Ball, above-mentioned, in 50000 Years.

Good God! cry'd she, how immense and wonderful are the Works of thy Hands! Why then, said she, if all the Stars were to be extinct or annihilated this next Night, we should not miss them till about 6 Months after!

No, Madam, said I; that Stream of Light now flowing from them to our Eyes,

Eyes, should the Fountain be stopped, would be half a Year before it would be run quite out; tho' it run after the rate of above 10 Millions of Miles in a Minute; a Motion almost as quick as Thought itself, as we usually say.

WELL, said she, this hath made an extravagant Notion of Mr. *Whiston*'s about the Distance of *Heaven*, or the Region appointed for the Bodies of the Blessed, its not being by any means so far off as the supposed *Empiræan Heaven* of the Divines, much easier to me than it at first appeared; and which then I thought a very new, wild, and unaccountable Opinion. But, pray, lets go on; Is not the *Number* of these fix'd Stars as wonderful as their *Distance*?

Vid. Acompt. of S. S. Prophecies, p. 288.

YES, in Truth is it, Madam, said I; for as the naked Eye discovers immense Numbers of them in a clear Night, (above 1000 of which are distinguished and taken notice of) and many more in the Northern cold Countries than we can do here; so when assisted by a Telescope of any great Length, it sees amazing Crowds of other Stars, which because they are invisible without these Helps, the Astronomers have called very properly *Tele-*

Telescopic Stars. *scopical Stars.* Such a Glass as this which we but now used to observe the Sun with, will discover to you many Thousands of Stars, before invisible to the naked Eye; and I think I have told above 70 within that little *Bunch of Stars*, which we call the *Pleiades*, or the 7 *Stars*; tho' now there appear but 6 to the bare Eye. *Milky Way.* The *Milky Way* is crowded with infinite Numbers of small Stars, from whence, as is usually thought, its *Whiteness* appears; which is a Discovery entirely owing to the Telescope; but whether the Whiteness proceeds from the *Smallness* of those numberless Stars, their *Nearness* to one another, or their *immense Distances*, we can't yet certainly determine, but must leave to Time and future Observations.

How endless is the Extent of the Divine Power and Goodness, said she, and how far are we yet from knowing the Bounds of the Starry World! But, Sir, your hinting, that formerly there were seven where now there appear but six Stars in the *Pleiades*, reminds me of your Promise, to acquaint me with the History of some Appearances of *New Stars*, and Disappearance of others, in *Cassiopæia*'s *Chair*, and in some other Places in the Heavens.

MADAM,

MADAM, ſaid I, the *Milky Way*, in which *Caſſiopæia* is placed, hath been famous for theſe Appearances; many new Stars having been diſcovered in the *Swan*, *Andromeda*, the *Ship*, *Eridanus*, and other Conſtellations within that Tract: Some of which have, after ſome Time, diſappeared, and then re-appeared again: Of theſe things you may ſee ſeveral Inſtances collected by the Author of *Lexicon Technicum* (a Book which your Ladyſhip hath) under the Title of *Fixed Stars*, in the ſecond Volume. But 'tis difficult to determine, what theſe new Stars are; ſome fancy them to be Planets revolving round ſome of the Stars in the *Galaxy*, and which therefore become viſible only in that Part of their Circle which is next to us; others take them to be Comets, and others think that they are really *fixed Stars*, whoſe Light and Vapours expire, but are again recruited and enkindled by the Acceſs of Comets towards them: But theſe *Hypotheſes* can't well ſolve all the Phænomena; for beſides the Appearances of theſe *New Stars*, it hath been obſerved of the known fixed Stars themſelves, that ſome of them have much changed their Magnitude and their Light; ſome of them have

Ricciol. Almageſt Hevelij. Prodrom. Mercator Aſtron. in Append. Philoſoph Tranſ. Miſcellania Berlin. Whiſton's Aſtron. Derham, *&c.*

quite disappeared for a Time, and then come into sight again; and this at certain Times and determined Intervals. And when you come to read what Mr. *Huygens* observed of the Stars in *Orion's Sword*, you will meet with what will very much, and I believe very agreeably, surprise you; but let it be which Way it will, 'tis a wonderful *Phænomenon*, and perhaps will never be thoroughly known, if ever, till future Ages have increased our Observations, and improved our Reasonings upon them.

I THINK, said the Lady, I have enough for this Time, about the Sun and the fixed Stars; I will consider of it, and have Recourse to the Books you recommend to me, and trouble you the next Time, about the Planets, in the Order as they are in, with Respect to the Sun; only give me leave to break in upon it, with regard to our own Planet, the *Earth*, and her Attendant, the *Moon*. With which, out of Self-love, or rather Inclination to the Place of our Birth and Abode, I would fain begin, if you don't judge it to be improper.

BY no Means, Madam, said I; for many things will occur in our Discourses about

about the Moon and Earth, which are very common and obvious Appearances, and which thoroughly accounted for and explained, will render the Knowledge of the other Planets much more easy and intelligible.

NOT long after this, the curious Lady attack'd me again, thus; I have been considering, said she, the amazing Subject we discoursed upon the last Time, and am prepared now to talk with you about the *Earth* and the *Moon*, and the different Magnitudes and Motions of each; and of this I find it previously necessary to have some Knowledge, or else my Enquiries into *those* of the other Planets, will not give me sufficient Satisfaction: Pray, Sir, how many Miles is the Diameter of our Earth reputed to be, by the Astronomers?

Something less than 8000, Madam, said I; and because I know you will expect it, I must tell you, that we attain this Discovery thus: Both in *England* and in *France*, a Measure in Length hath been taken upon the Earth's Surface, under one and the same Meridian, or, in a right Line running exactly North and South, till by accurate Instruments it was found, that the *Pole* was *raiſed* or *depreſſed* exactly one Degree. This, the Mathematicians of both Nations agreed in to be almoſt 70 Miles, *Engliſh*. And there being 360 Miles in a Degree, that Number, multiplied by the former, gives you the Number of Miles in the whole Circumference of a great Circle on the Earth, or how many Miles it is round our Globe; and then, by the Principles of Geometry, they know, that ſomething more than one Third of that muſt be the Earth's Diameter. I don't trouble you, Madam, with the exact Numbers, nor the Multiplication and Diviſion, but you may depend upon it, that the round Number of 8000 Miles, is pretty nearly the Earth's Diameter, tho' ſomething too much: And the half of this, *viz.* 4000 Miles, is the *Semidiameter*, or the Diſtance from the Surface to the Earth's Centre, a Number, or Meaſure, much uſed by Aſtronomers.

Earth's Semidiameter.

I THANK you, Sir, quoth ſhe, for this; the Knowledge of this ſingle Point, will I ſee carry me a great Way, when I come to read Aſtronomical Authors; But, pray, Sir, go on and oblige the with a further Account of this Earth; Do you think it really turns round its

Axis,

Axis, as you have found the Sun to do by its *Spots*?

YES, Madam, ſaid I, and as there is nothing more eaſy and ſimple than this Motion, ſo it accounts for the Appearances of Day and Night in an eaſy and natural Manner; for as this Earth revolves from Weſt to Eaſt in exactly 24 Hours Time, it makes the Sun *appear* to do ſo from Eaſt to Weſt in the ſame Time; and makes it *Day* to thoſe Places on its Surface, which are turned towards the Sun, and *Night* to ſuch as are in the oppoſite Parts; as you ſee, Madam, if I ſet this Globe into the Sun's Light, it will illuminate *but one Half* of it, and the *other Half* will lie in the Shadow; but as I turn the Globe round its Axis, all Parts of the Earth's Surface painted upon it, will come ſucceſſively into the *Light*, as the oppoſite Parts go, after the ſame Manner into the *Dark*.

Cauſe of Day and Night.

I GRANT you, Sir, ſaid ſhe, this is a very natural and eaſy Way of accounting for the Viciſſitudes of Day and Night; and ſo ſhort and unembaraſſed in compariſon of the other wild Notion, which makes the Sun, and all the Region of the fixed Stars, to revolve round us in 24 Hours, that

that it recommends itself to us, at first Sight, as agreeable to the other Proceedings of Nature, if we could but get rid of our Prejudices, so as to conceive it possible to be done, without our perceiving it. But can we travel above 1000 Miles in an Hour, and not be sensible of it?

As easily as 10 in a Ship, Madam, said I; where, let the Vessel move never so fast forwards, if it were not for the Tossings and Shocks which the Resistance of the Water and Waves make, and for the Rustling and Bustle that the Wind makes in the Sails, you would perceive no Motion at all in the Ship, but judge it to be perfectly at Rest; and if another Ship lay at Anchor by you, you would judge that to move *backwards*, and not your self *forwards*. And much more will this appear plain, if you, consider that with the Earth's Motion round its Axis, the Air, and all the Atmosphere moves along with you, and doth not resist you, as is the Case in the Motion of a Ship. But indeed, the greatest Wonder in this Case is, that we are not all whirl'd off into the Air, like Dirt from a Wheel, or Drops of Water from a twirling Mop, or Stones part ing from a Sling.

YOUR

YOUR talking of the *Twirl of a Mop*, said she, puts me in mind of a whimsical Description of *that Action*, which a Friend of yours made to ridicule some *Verbose Verses* then repeated: But tho' I have almost forgot them, I hope you have not.

MADAM, said I, your Ladyship's thinking of them now is proper enough; for tho' made to *expose* another Matter, they will *illustrate* what we are upon:

See how Culina *with hard adverse Wrists,*
The dreary Radii of her Mop *untwists;*
Swift twirling round, the oblong Planet *rolls,*
With Axe produc'd thro' the Meridian Poles;
The Stiff'ning Threads their rigid Form preserve,
While dirty Drops fly off in Tangents to the (Curve.

WHY this is very true, said she, of those dirty Drops, and I can't imagine why 'tis not so with us; for I don't know any thing that fastens us down to the Earth, but our firm Inclinations to this World, which I believe yet hath no Physical Power to keep our Bodies annexed to its Surface. Pray, how do you account for this Difficulty?

Centripetal Force.

BY

BY that Will of the Creator, Madam, which we call the *Law of Gravity*, or *Gravitation*; whereby all heavy Bodies have a Tendency towards the Centre of our Earth, in such an over Proportion, that the *Centripital Force*, by which Bodies tend thither, is almost 300 times greater, than that by which they are forced off by the Earth's Motion round its *Axis*, or the *Centrifugal Force*, as they call it; and 'tis this *All-wise Provision* that keeps all things together on the Surface of the Earth; and which, when exactly adjusted, keeps also every Planet in its proper Circle, and at its due Distance from the Sun, or from its Primary one: And this is so universal a Law, that it prevails every where: And if a Cannon-Ball could be discharged from any considerable Height, in the Air, parallel to our Horizon, and with a Velocity equal to that of the Earth's Attraction, or the Force of Gravity towards the Earth's Centre, it would then neither *fall* to the Earth at all, nor go quite *off* from it, but would *revolve round* it, like our Moon; and this is the very Reason why she doth so.

WELL, said she, a new World of Knowledge opens and dawns upon me! I begin

begin to ſee a thouſand Things, of which I had no Notion before; and I believe the Motions of the heavenly Bodies, after this, won't appear ſuch abſtruſe unintelligible Things as I imagined them to be: But, pray, Sir, explain this a little further, with regard to the Moon.

YOU muſt know, Madam, ſaid I, that this Gravitation of a Planet towards any Central Body, decreaſes vaſtly, as the Diſtance from that Centre increaſes; and therefore the Moon being about 60 of the Earth's *Semidiameters*, or 240000 Miles diſtant from us; her Gravitation towards the Earth, will be 3600 time leſs than that of a Cannon-Ball ſhot out of a Gun on or near the Surface of our Globe; and the great Creator hath ſo wonderfully contrived it, that her *Centrifugal* Force, or her Endeavour to fly off from the Earth, is exactly equal to her Gravitation thither; and this keeps her in her Orbit, as it doth all the Planets in theirs, as I ſaid before.

Moon's Diſtance

O wonderful and happy Adjuſtment! ſaid ſhe, for I perceive, if the Moon's Gravity towards the Earth were much abated, ſhe would run out of her Orbit, and leave us; and if the oppoſite Force were

were much lessen'd, she would, in a little time, tumble down upon us: Am I right, Sir, in this Conclusion?

EXACTLY, Madam, said I, and I perceive I need not say much more to you upon this Head, except it be to tell you, that if the Centrifugal Force were taken away from the Planets, and that only the Power of Gravitation towards the Sun remain'd, they would all soon fall down to him, and our Earth would get down thither in about 64 Days and 10 Hours time.

But I think, Madam, we are gotten to the Moon a little too soon, having not yet quite done with the Earth, whose Annual Motion round the Sun therefore, let us next consider: By which all Increase and Decrease of Day and Night, and the *Changes* and *Seasons* of the Year are made.

AND can you give me any good Reasons, that I can understand, to believe this Annual Motion of the Earth, said she?

I THINK, Madam, said I, there is in Astronomy a *plain Demonstration* for the Motion of the Earth round the Sun; but it will be too remote for your present Knowledge of these Matters: However, I think

I think 'tis a very *good Argument* for its being ſo, that *this Way* there is a Parity and Agreement with the other Proceedings of Nature, which is very ſuitable to the Wiſdom, Eaſineſs and Conciſeneſs obſerved by the Divine Being: For it being now agreed, that the Sun is the Centre of all the other Primary Planets, and that we are placed in ſuch a due Diſtance from the Sun, between the Orbits of *Venus* and *Mars*, as anſwers to the *Time* of thoſe Planets Revolution round the Sun; and ſince 'tis alſo agreed, that the *other primary Planets*, as well as *Mars*, *Venus*, and *Mercury*, do, in their ſeveral Orbits, revolve round the Sun; what Reaſon can poſſibly be aſſigned, why the *Earth* ſhould not do ſo too? ſince *they are Earths* likewiſe as our *Planet* and the *Moon* are, and conſequently our Earth muſt be as capable of moving round the Sun, or any other Centre, as *they* or *ſhe* are.

I OWN, ſaid ſhe, that 'tis much more natural, orderly and harmonious to ſuppoſe it ſo, and therefore I will lay aſide all Prejudice, and believe it with a good Aſtronomical Faith.

MADAM, ſaid I, if you will obſerve what *Fontenelle* ſaith, very juſtly, of Nature,

ture, that *She is always magnificent in the Deſign, but frugal in the Execution of it*, you will never believe that the Sun and fixed Stars turn all round us in 24 Hours; when you reflect, that the bare Motion of the Earth round its Axis will anſwer all your Ends that are to be ſerved by the other. That would be juſt as abſurd, as for a great Architect to contrive, with vaſt Expence and Machinery, a *Kitchen-Grate*, that ſhould revolve round a Spit, in order to roaſt a *Wheatear* or a *Wren*; but never ſo much as dream of a Way to turn the Spit round.

'Tis monſtrous, ſaid ſhe, as well as ridiculous, and as I told you before, I won't believe one Word about it: I ſee, that the more *plain* and intelligible things are, the more they are valuable; and that *Obſcurity* and *Myſtery* are uſually the Effects of Ignorance, and want of Skill either in the Operator, or the Explainer. But, Sir, will you give me leave then to ſtep to the Moon, and aſk you a few Queſtions about her, for I can't put thoſe fine Lines of *Butler* out of my Head:

Moon.

The Moon put off her Veil *of Light,*
Which hides her Face by Day from Sight;

Myſte-

Myſterious *Veil of* Brightneſs *made*,
That's both her Luſtre *and her* Shade;
And then indeed as freely ſhone,
As if her Rays had been her own:
For Darkneſs is the proper Sphere,
For borrow'd *Glories to appear.*

And I know a good deal of his Meaning in them; as *that the Moon borrows her Light from the Sun*, &c. but I could be glad if you would explain a little upon that Matter, and upon her Motion round the Earth; after we have at Night examined her Face by the Teleſcope: Is this a good Time to look at her?

YES, Madam, ſaid I, a very lucky one, for ſhe is now increaſing, and not quite full; we ſhall ſee her *Mountains* more diſtinctly, and the Light of the Sun move from one Hill to another.

THE Evening, according to our Wiſhes, proved very clear and fair, and the Lady was mightily pleaſed with the *Face* and Appearance of the Moon thro' the Glaſs; and having alſo the Day before been reading a little in Mr. *Huygens*'s *Coſmotheoros*, or his Celeſtial Worlds diſcover'd, or Conjectures about Worlds in the Planets; and in Mr. *Whiſton*'s late

Book, called *Astronomical Principles of Religion*; she was prepared to ask me some very proper Questions, and began thus:

PRAY, Sir, said she, is not our Earth a *Moon* to the Lunar People, as well as *she* is to ours?

YES, Madam, said I, and a most useful and a glorious one too; and we may in some Measure perceive *that* our selves, by the Light which our Earth reflects upon the Moon before she is just *new*, and for some Time after; for doubtless *that* is the *only* Light that then renders her visible to us: And when you consider that the Light of our Earth, consider'd as a *Moon*, will be thirteen times greater than that of the *Lunar Light* to us, it won't appear strange, that its Reflection on the Moon should render her *then* dark Body visible to us. However, this Terrestrial Light, when the Earth appears at Full to the People in the Moon, is not above a 3600th Part of the Sun's Light there, as the Light of our Full-moon to that of the Sun, shining upon us, is about As 1 To 48000.

I THANK

I THANK you for this, Sir, ſaid ſhe, and am heartily glad *we* can be ſo uſeful to the *Lunar World*. But, pray, go on: I perceive, ſaid ſhe, you agree that the Moon as well as all the reſt of the Planets, turns round her own Axis, which, methinks, in her is very ſtrange; for we ſeem to diſcern always the ſame Face of her, without any Variation: Pray in what Time is that Motion perform'd?

JUST in the Time, Madam, ſaid I, that ſhe is revolving round the Earth; which I will explain to you preſently: But, firſt, it will be proper to inform you, that the Figure of the Moon not being exactly globular or ſpherical, but a little oval, or like that of an Egg, her longeſt Diameter (which exceeds her ſhorteſt by about 200 Foot) would, if you ſuppoſe it extended ſo far, paſs thro' the Centre of our Earth: And hence it is that we ſee always the ſame Face of the Moon obverted towards us, and that *this* is not hinder'd by her Motion round her Axis, this familiar Inſtance will ſhew you. Pleaſe to ſit ſtill, without turning your ſelf, while I walk round you; you will then ſee plainly, that if I keep my Face always towards one and the ſame Point of the Compaſs, while I am moving round you, when I come a Quarter of my Circle, my right Side, and not my Face, will be towards you; when I have gone half Way round you, my Back; and when I'm gotten three Quarters, my left Side will be turned towards you; but if, as I move in my *proper Orbit* round you, I always keep turning towards you, as indeed I can't help doing, you will then always obſerve me beholding you with the ſame Face of Reſpect and Eſteem.

YOU Men, ſaid ſhe, are not like thoſe conſtant celeſtial Lovers; for you ſeldom continue your Reſpect for above a Revolution or two: however, you may now ſtop in your *Circular March*, continued ſhe, for I ſee the thing plain, and that the Reaſon why we ſee always the ſame Face of the Moon, is becauſe ſhe moves round her *Axis* in the ſame Time that ſhe performs her Circle round the Earth: But, pray, let me know ſomething more of the Manner of her Motions.

MADAM, ſaid I, the Moon revolves continually from Weſt to Eaſt, and that pretty nearly in the ſame Circle which we call the Ecliptick; but not exactly ſo, ſometimes running 5 or 6 Degrees above it it to the North, and ſometimes below it to the Southward: She doth not alſo keep always the ſame Diſtance from the Earth; as appears by her Diameter, which when we come to meaſure, we find ſometimes conſiderably larger than at others; ſhe moves ſwifter in the *Syzygys*, as they call them, that is in their *Conjunctions* with and *Oppoſitions* to the Sun, than ſhe doth at her *Quadratures*, or when ſhe ſhews juſt half of her Face.

WELL, ſaid ſhe, I perceive now that her Motion is ſo irregular, that ſome Compariſons, which have been made with her, are not quite groundleſs: But this Part I fancy I ſhall get over by my Books, and I think I know alſo, that the Reaſon why ſhe appears *full*, is, becauſe ſhe is then *oppoſite* to the Sun, who ſhines *full* upon her; and we loſe Sight of her in what we call the *New Moon*, becauſe ſhe is then between us and the Sun, or in Conjunction with him; and 'tis eaſy to ſee alſo that all her other *Changes* and *Appearances*, or *Phaſes*, as I remember you call them, are accountable from her being in ſome *intermediate Poſition* between *new* and *full*. But, pray Sir, why have not we an Eclipſe of the Sun at every *new* Moon, and one of the Moon at every *Full?*

THAT is owing, Madam, ſaid I, to the Moons Latitude, by which ſhe runs ſometimes 5 or 6 Degrees from the *Ecliptick*, (in which the Earth always moves) both Northward and Southward. But if her Orbit, and that of the Ecliptick, were all in one Plane, there would be total and central Eclipſes at every *new* and *full* Moon.

I CONCEIVE what you ſay, ſaid ſhe; ſo that there can be no *Eclipſe* of either Sun or Moon, unleſs the Moon be in the Ecliptick as well as the Earth, becauſe the Sun's Light will go *by* or *beſides* the Earth or Moon.

YOU have it exactly right, ſaid I, Madam, in general; all that I need tell you further is, that if the Moon have but a *very little* Latitude an Eclipſe may happen; or if ſhe be at the Time of the *Conjunction* with, or *Oppoſition* to the Earth, in or near the *Nodes*, as they call it, that is, the Points where the Moon's *preſent* monthly Circle croſſes the Ecliptick. And this falling out commonly twice in every *Synodical* Month, or *Lunation*, as we call it, there would be an Eclipſe of the Sun and

and Moon both, if the Earth could ſtay about the *Nodes*, and did not proceed on in her Orbit all this while, or change her Place in the Ecliptick forward on. However, within the Compaſs of the Year, there happen uſually four notable and almoſt total Eclipſes, ſomewhere or other, two of the Sun and two of the Moon. But your Ladyſhip will pleaſe to conſider, that there is in the Nature of the thing a great deal of Difference between theſe two kind of Eclipſes: In the *Lunar Eclipſe* there is a *real Loſs* of the Moon's Light, and it is alſo in the *whole* the ſame, from whenceſoever it is ſeen, not being changed by the diverſe Poſition of the Spectator on any Part of the Earth's Surface, whether he be in the Equator, or at the Poles.

But in an *Eclipſe of the Sun*, there being no real Loſs of the Sun's Light, but only an Interception of part of it, from coming to our Eyes, by the Interpoſition of the Moon's Body; this Eclipſe muſt appear different according to the different Places on the Earth, from which a Spectator may obſerve it; for tho' to that part of the Earth to which the Centre of the Moon is then interpoſed between the Obſerver's Eye, and the

Fig. III. Sun's Centre, it will be *total*, yet it will be but a *partial* one to all other Places, and none at all to the remote ones, as you will eaſily ſee by *Plate* III.

I THANK you for explaining the *Nodes* to me, ſaid ſhe, and theſe Phænomena of Eclipſes; and by it you have now ſaved your ſelf the Trouble of anſwering many Queſtions: And I perceive, for the future, I ſhall get to underſtand theſe kind of Things and their Terms of Art, pretty well: My *Lexicon Technicum* is a ready Help to me in time of need, and I believe the Doctor compoſed it out of a peculiar Regard to our Sex; I'm ſure we are very much obliged to him for it. But have we any thing further to ſay about this *Vagrant*, the Moon, of whom *Dryden* ſpeaks a little coarſely, methinks, in his Trànſlation of *Ovid:*

Nor equal Light the unequal Moon adorns,
Or in her wexing, or her waneing Horns;
For every Day ſhe wanes, her Face is leſs,
But gathering into Globe, ſhe fattens at Increaſe.

MADAM, ſaid I, happy is your Taſt in every thing! A piercing Judgment, great Memory, ſedate Conſideration, and fine Luxuriances of Wit, ſeldom unite in

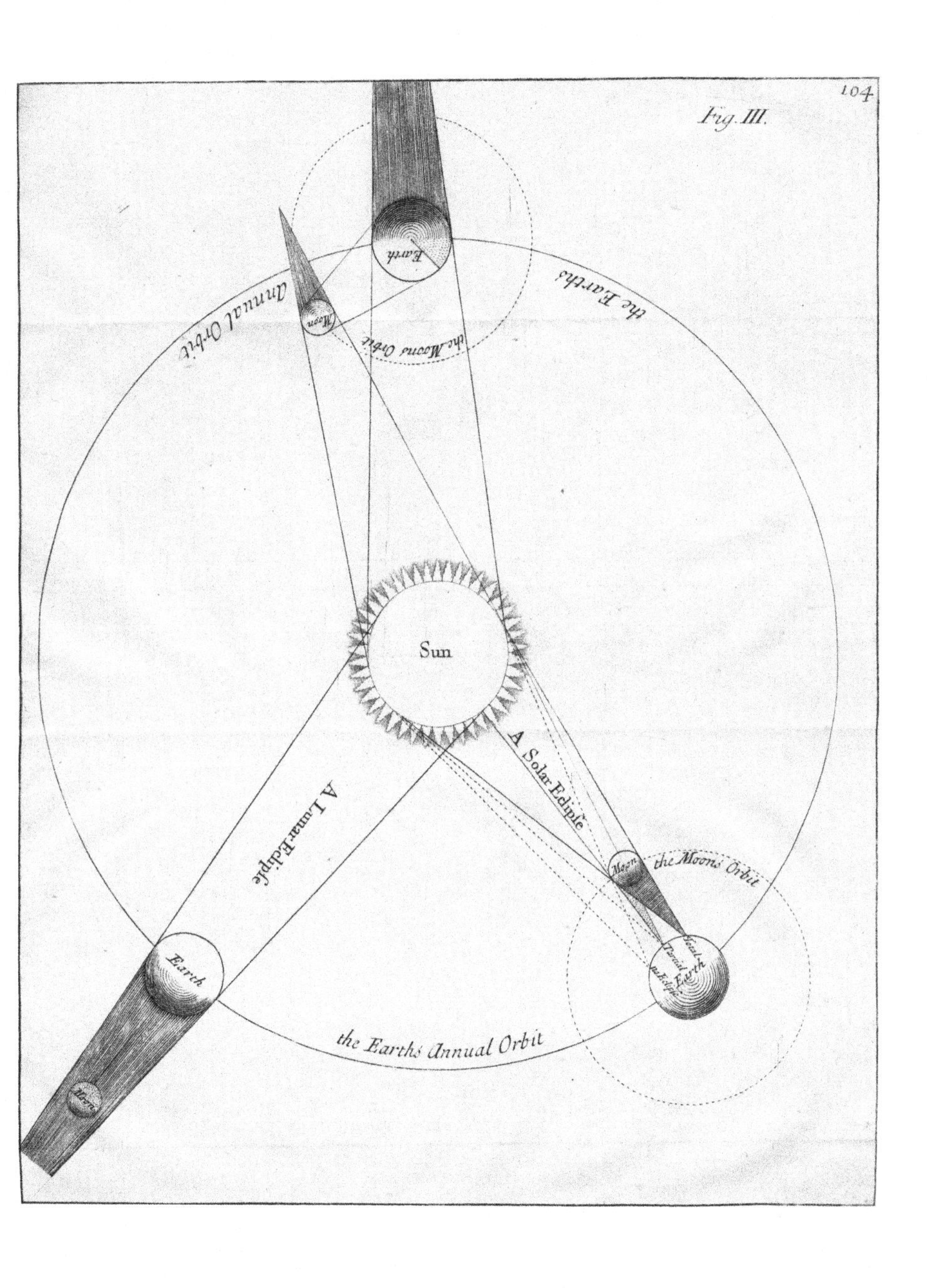

The material originally positioned here is too large for reproduction in this reissue. A PDF can be downloaded from the web address given on page iv of this book, by clicking on 'Resources Available'.

in one Perſon; but you are adorn'd with all.

PRAY, Sir, ſaid ſhe, ceaſe your high Strains, or elſe I ſhall begin to think, as *Hudibras* expreſſes it, That

The Queen of Night, whoſe vaſt Command,
Rules all the Sea and half the Land;
And over moiſt and crazy Brains,
In high Spring-Tides at Midnight reigns,

had laſt Night ſome Effect on my *Teacher*'s Head.

MADAM, ſaid I, between you both, I own I am now and then a little flutter'd. ——But your laſt Verſes mentioning the *Tides*, put me in Mind of another great Influence and Uſe the Moon hath over this Earth, beſides the great Light ſhe gives us; and that is, that ſhe is the principal Cauſe of our *Tides*, which are ſo beneficial to us, in keeping our Ocean ſweet by their Motion; and in helping the Navigation of our Ships in Rivers, and Places of the Sea, near the Shores. But I will refer you to the *Lexicon Technicum*, under the Word *Tides*, for an Account of it, where you may receive full Satiſfaction, with regard to this Affair; *formerly*

merly ſo very unaccountable, but *now* very clear and intelligible.

I CAN'T leave, ſaid ſhe, this inconſtant Planet, whom your ill-natur'd Wits have ſo often made the Reſemblance of our Sex, without asking you a Queſtion or two more; What is that you call the *Hunter*'s *Moon?* I have heard the jolly Fox-Chacers talk much about it.

MADAM, ſaid I, what they mean by it, I don't underſtand; but I ſuppoſe it muſt be ſome very *long Moon*, which ſhines a great while about the hunting Seaſon, and ſo hath become eminent that Way, tho' this can't always ſo happen: But if you will pleaſe to go to the Globe yonder, you will eaſily ſee, that when the Moon is in *Cancer*, her *Sweep*, or the Ark which ſhe makes above our Horizon, will be vaſtly larger than when ſhe is on the other oppoſite Side of the Equator in *Capricorn*; and if it then happen to be *full Moon*, or pretty near it, (as will be the Caſe next *October*) the enlighten'd Ark of her Motion above the Horizon will be very large and conſpicuous, in compariſon of what ſhe will run when her *full* ſhall happen to be in *Capricorn*. But the Moon's Lati-

Latitude won't much help to account for this Phænomenon; for *that* is greater at her *Quadratures*, *i. e.* when ſhe appears an Half-Moon, than at the *Syzygies*, as we call them, *i. e.* at *New* or *Full Moon*; or when the Sun, Moon, and Earth are all nearly in one Right Line.

I THINK I apprehend this, ſaid the Lady; and now for the reſt of my Queſtions. What is the *Diameter* and Magnitude of this Planet? What is her Motion? What Proportion doth ſhe bear to our Earth? And do you think ſhe is inhabited as we are?

HER Diameter, Madam, ſaid I, we reckon to be almoſt 2200 Miles; in Degrees, when taken with an Aſtronomical Inſtrument, it is about 32 Min. 12 Seconds, which is nearly the ſame with the Sun's apparent Diameter, for that is but 31 Min. 27 Seconds; her Magnitude, or rather her *Maſs*, or the quantity of Matter in her, with regard to the Earth, is about $\frac{1}{40}$ part; but then the Denſity of her to that of the Earth, Is as 9 To 5; ſo that if ſhe hath any Inhabitants, as I take it to be highly probable all the Planets have, they can't be of *ſuch Conſtitutions* as we are

And

And tho' some of our Astromomers have lately concluded from some Observations made in Eclipses of the Sun, that she hath an *Atmosphere*, or Air about her like our Earth; yet 'tis probably of a very different Nature from ours, without any Clouds, Rain, Hail, or Snow; because, whenever our Air is clear, we can always discern the Moon's Face, with, as well as without a Telescope, to be bright, clear and distinct: Which I think could not well be, if her Atmosphere were like ours.

The exact time of her Periodical Revolution round the Earth, is in 27 Days, 7 Hours, and 43 Minutes, and this is call'd her *Periodical Month*; in which Course she runs about 2200 Miles in an Hour. But her *Synodical* one, as they call it, or the Time from *New Moon* to *New Moon*, is 29 Days, 12 Hours, and $\frac{3}{4}$ of an Hour.

PRAY, said the Lady, what occasions this Difference of above 2 Days and 5 Hours, between these two kinds of *Lunar Months?*

THE Reason, Madam, said I, you will easily apprehend; and 'tis this: While the Moon is revolving round the Earth in her

her *Periodick* Month, the Earth it ſelf is moved on in her Orbit round the Sun almoſt an entire Sign, or one twelfth part of the *Ecliptick* : and therefore *that Point* in the Moon's Circle or Orbit, where the laſt Conjunction with the Sun was made, will now be gotten too far to the *Weſtward:* and therefore ſhe cannot come again to a Conjunction with the Sun 'till after 2 Days and about 5 Hours ; which Time muſt be paſs'd before the Moon can have exhibited all her *Phaſes.*

I hope, ſaid ſhe, I ſhall get to conceive this a little better by degrees ; but pray let me go on now ; and ask you a Queſtion or two more : I have been thinking, that the Inhabitants of the Moon muſt have one thing very odd and ſtrange ; and that is, that to one half only of their World *our Earth*, which I am appriſed muſt appear as a *Moon* to them, can be viſible : So that their other Hemiſphere will be for ever deprived of the Advantage of a Moon's Light.

O! Madam, ſaid I, if your Speculations lead you into ſuch Depths, we have you ſafe for an Aſtronomer ; and I don't doubt but that will lead you alſo into the Study of ſuch other Parts of Mathema-

ticks

ticks as, the more you know of them, the more you will find them necessary. And I could now tell you a great many surprising things about the Appearance of our Earth to the Inhabitants of the Moon; but I will not deprive you of the Pleasure of reading them your self: you will find them fully enlarged upon at the End of Dr. *Gregory*'s *Astronomy*, which is lately translated into *English*; which you will find among those Books, that, according to your Commands, my Bookseller sent you last Night from *London*.

VERY well, Sir, said she; I shall be impatient till I get some further Knowledge of that Matter. But we will now take our Leave of Mrs. Moon; and, if you please, go *down*, as you call it, towards the Sun: and from thence ascend again; taking the rest of the Planets in their Order, according as this Diagram here represents them; which you have kindly drawn for me; and which you call a Scheme of the *Solar System*. Pray therefore, good Sir, tell me as much as you think I can understand, about *Mercury*, the nearest Planet to the Sun.

See Fig. IV.

Mercury.

MERCURY, Madam, said I, is a Planet whose Diameter we reckon to be about

about 2700 miles ; and therefore he is about two thirds of the Earth's Magnitude. His Diſtance from the Sun is about 32 millions of miles ; and his mean Diſtance from us, about 22000 of the Earth's Semidiameter, or 88000000 miles, according to *Caſſini*'s Numbers. He revolves round the Sun in ſomething leſs than 88 days, with the Velocity of 100000 miles in an hour : which is almoſt as faſt again as the Earth travels : for we don't go above 56000 miles in that time ; and yet that is making pretty good ſpeed too ; for that don't want much of a 1000 miles in a Minute, or 15 miles in a Second ; or in that ſpace of Time in which you can diſtinctly pronounce *one*, *two*, *three*, *four*. And yet however amazingly ſwift this may ſeem, 'tis crawling like the *American Ignavus*, or Beaſt called the *Sluggard*, in compariſon of the Velocity of the Rays of Light, which certainly move about 180000 miles in that Time.

What! in a Second? ſaid ſhe : Let me ſee— ; why, that is almoſt 50000 miles while I can ſay the word *Light*. For godſake ſtop a little, or you will make me perfectly giddy : my Head will turn quite round What! have you and I then been travelling almoſt 2000 miles together this

this Morning, and I knew nothing of the matter ?

'Tis even ſo, Madam, ſaid I ; and you ſee we move eaſily : But if you pleaſe I will go on. The Heat of the Sun there, is probably 7 times (Mr. *Huygens* ſaith 9 times) as great as with us in the hotteſt Summer ; which is, I believe, enough to make Water to boil. You will eaſily ſee therefore that his *Inhabitants* cannot be *ſuch* as we are ; for our Bodies could by no means bear ſuch a Degree of Heat.

OUR *Anceſtors* Bodies, ſaid ſhe, I believe could not : but by our drinking ſo much *ſcalding Tea* and *Coffee* as we now do, I ſhould think we are preparing ourſelves to go and live there : And I ſuppoſe our famous *Fire-Eater* came from thence. There can be no Fluids ſure in this Fiery Planet, much leſs Denſe, than *that* which bears his Name ; and no doubt all things elſe are Denſe there in the ſame Proportion, or elſe the Sun would rarify him, and ſend all his Furniture off in Fume, Smoak and Vapour.

Well ! ſaid the Lady, as much as I hate frozen Zones and bitter cold Weather, I think this Mercurial World to be worſe in the other Extream ; ſo I will never wiſh for a Voyage thither. No,

No Madam, said I, you will find this Earth to be a much more Eligible Place of Abode for People of *our make*, than any other which we yet have discover'd in the best Planet of them all. As for this we are talking of; *Mercury* is so near the Sun, that he is very rarely seen by any but Astronomers, who know how to look after him. But about St. *George*'s Day last he was at his greatest Distance from the Sun, and then about 8 in the Evening might have been seen very plainly.

WELL, said she, I shall not much trouble myself to enquire after him; but I remember a saw him very plain and distinct, during the Total Darkness, in the last Eclipse of the Sun; and that shall satisfy my Curiosity, till some other Opportunity offers it self. But pray Sir, doth the Telescope shew us any thing remarkable about him?

ONLY Madam, that he hath *Phases*, as we call them, like those of the Moon, and sometimes appears full, and sometimes horned, like her; which you will easily conceive must be the Case of any Globe of Earth illuminated by moving round the Sun, and changing its Position, with regard to him, and to our Eyes. It hath not been yet discovered by any Spots or Marks upon him, that he revolves round his *Axis*, nor consequently what the *Position* of that *Axis* is, tho' 'tis probable he performs that Motion in a certain and determinate Time, as the rest of the Primary, and I believe all the Secondary Planets do. *Venus* and our *Earth* must needs appear very bright and large to the Inhabitants in this Planet, and the former will seem 6 or 7 times larger than she doth to us, which will help to supply the want of a Moon to him in the Night. But there is one more very remarkable Phænomenon of him, and that is, that as his Orbit is within ours, he must sometimes get between us and the Sun, and then he appears like a little black Spot in the Face of that Luminary, and may very easily be observed and distinguish'd by a Telescope.

O, I am mightily pleased with this, said the Lady, and shall I ever see him in that Position?

I Hope you will many a time, Madam, said I, for he will be there in *April* 1720, and in *October* in 1723, which is is but a little while hence; and he will alſo be there again in *May* 1761.

WELL, ſaid ſhe, I will then have a full look at him, if I live ſo long; and in the mean time let this *Herald* of the Gods ramble on as he pleaſes; and let us talk next about *Venus*.

Venus.

Beneath the ſliding Sun, who runs her Race,
Doth faireſt ſhine, and beſt become the Place:
For her the Winds their Eaſtern Blaſts forbear,
Her Month reveals the Spring, and opens all the (Year.
With ſmiling Aſpect ſhe ſerenely moves!
Adorns with Flowers the Meads, with Leaves (the Groves.
The joyous Birds her Welcome firſt expreſs,
Whoſe Native Songs her Genial Fire confeſs.

Dryden's Lucretius.

But whither am I running? Pray Sir, ſtop me a little, and tell me ſome ſerious Aſtronomical Things about this celebrated Planet.

THE Diſtance of *Venus*, ſaid I, Madam, from the Sun is about 60 Millions of Miles; and by ſome *Spots* which the Teleſcope hath diſcovered in her Face, ſhe appears to have a Revolution round her *Axis*: The Time of which ſeems to be about 23 Hours. But neither *Cassini* at *Paris*, nor our Mr. *Hook* here, tho' they plainly saw the Spots to move, were able, positively and expresly to determine the Time of her Diurnal Rotation round her *Axis*; tho' the former takes it as I said before, to be in about 23 Hours; and therefore *that* will be the Length of her Natural Day. Her Motion in her Orbit round the Sun, is performed in a little above 224 Days, and her Motion in an Hour is about 70000 Miles.

THAT's pretty fair, said she, too for a Lady; but I am glad she doth not fly quite so fast as the last *Whirlegig Mercury*, however: But pray Sir, go on.

THIS Planet, Madam, said I, Mr. *Huygens* takes to have a large Atmosphere, which reflects so strong and glaring a Light, that her Body is rarely seen clear and distinct. She also hath *Phases* like the Moon; as was before observed of *Mercury*; she hath no *Satellites*, Attendants, Moons, or Secondary Planets moving round her, because as you very justly observed a while ago, *Mercury* and she being so near the Sun, have no occasion to be enlightned by *Moons*, as our Earth, *Jupiter* and *Saturn* have. Indeed *Cassini*, in the Years

1672,

1672, and 1686, with a Teleſcope of 34 Feet, fancied he ſaw a *Satellite* moving round her, whoſe Diameter was about a quarter part of that of *Venus:* And Dr. *Gregory* thinks it not improbable, that this might be really a Moon to this Planet, and takes the reaſon of its not being uſually ſeen, to be, the unfitneſs of its Surface to reflect the Rayes of Light: But as no ſubſequent Obſervations have confirmed this, I look upon it no more than a Conjecture. Neither ſhe nor *Mercury* ever come ſo much as into *Quadrature* with the Sun, much leſs to an Oppoſition to him; and indeed, their utmoſt *Elongation* from him, as we call it, or greateſt Diſtance Eaſt or Weſt from the Sun, never amounts to above 2 Signs; *Mercury* not going above 28, and *Venus* never above 48 Degrees from the Sun. She is much about 40 times larger than our Earth; if, as ſome ſay, her Diameter be 7 times as long as that of our Planet: And the Light and Heat of the Sun, is about 4 Times as great as it is with us.

I'M heartily ſorry, ſaid the Lady, that 'tis ſo; for I would fain have had this beautiful Planet to have been inhabited by juſt ſuch fine Gentlemen and Ladies as we have here; but I find 'twont do;

the Women wou'd be there all as ſwarthy as Gibſies, and fry and ſweat like Negroes in *Africa:* Out upon it! I'm afraid I ſhall find never a Planet fit to be inhabited by ſuch People as you and I are.

MADAM, ſaid I, take Care; you are falling in with the Aſtrologic Whimſies; one would think you had read *Athan. Kircher*'s *Iter Extaticum*, which agrees with your Wiſhes as to *Venus*, *Mercury*, and *Jupiter*; but he makes *Mars* all Smoke and Fire, and *Saturn* nothing but dull Lead, Dirt and Naſtineſs, as you will find when you come to look over Mr. *Huygen*'s Planatary Worlds, which I have ordered the Bookſeller to ſend you.

YOU are always cautioning me againſt Aſtrology, ſaid ſhe, and I muſt thank you for it. But I have heard that their *beginning* with that Study, hath made ſome Men become good Mathematicians, and even Aſtronomers: Shall I name them to you, Sir; you have forgot what you have told me of ſome of your Friends. But enough, let us proceed, and before we have quite done with this warm Dame, will you pleaſe to tell me, why ſhe is ſometimes our *Morning*, and ſometimes our *Evening Star?*

THAT

THAT depends Madam, ſaid I, on her Poſition, with regard to the Sun and us; when ſhe is in that part of her *Orbit* which is below the Sun, or between him and us, then ſhe is the *Morning Star*; but when ſhe gets into the oppoſite part of her *Orbit* above the Sun, then ſhe becomes our *Evening Star*.

AND under both thoſe Denominations, ſaid ſhe, I think the Poets make her change her Sex, and turn *He-Thing*, as if ſhe could not be as uſeful when of *our Gender*, as of *yours*; for thus, forſooth, Mr. *Dryden* Compliments the *Changling:*

So from the Seas exerts his Radiant Head,
That Star, by whom the Lights of Heaven are led,
Shakes from his Roſie Locks the pearly Dews,
Diſpels the Darkneſs and the Day renews.

And ſo that blind Creature *Milton* cries,

Bright Herſperus *that leads the ſtarry Train,* &c.

Marry come up indeed! Can nothing but Men ſerve you? Sure we have had Women every way as well qualify'd to be Morning or Evening Stars as any bearded Tyrant of you all.

MADAM, ſaid I, this is only owing to Cuſtom, which hath made it the Mens

Province to write Books and make Verses, and so they Compliment themselves: But however, you may be pretty easy, when you reflect, that we usually call the *most useful* things *She's:* Our *Saxon* Ancestors and our plain honest Country Folk, now call the *Sun* himself, that Father and Governor of all the Planets, *She*; and so we agree to call Guns and Fowling Pieces; nay, our Sailors are so well bred, and such Lovers of your Sex, that they call a Ship *She*, tho' she be a *Man of War*.

WELL! said she, this is some kind of Atonement and Satisfaction; and therefore at your desire, I will for this Time forgive the Gossips of *Phosphorus* and *Hesper*; but if they should attempt to make a *Man of the Moon*, I will never pass it by, for I can hardly be reconciled to those that place a *Man* in that Planet. But have you any thing further to tell me about *Venus?*

Venus in the *Sun.*

ONLY, Madam, that *She* also sometimes, like her Neighbour *Mercury*, hath appeared like a Spot in the Sun; as you will easily conceive she may, when you consider that the Orbit of the Earth includes her's within it; and that therefore she must be sometimes, tho' very seldom, between

between our Eye and the Sun, and then ſhe will appear like a Spot in the Sun's Disk. The next time that *Venus* will be ſeen in the Sun, will be *May* 26. 1761. a little before 6 in the Morning; I wiſh your Ladyſhip Health and Happineſs till the Time of that Obſervation, and that you may be then well enough to *get up* to ſee it.

O! Sir, ſaid ſhe, I can riſe betimes in a Morning, for a leſſer Occaſion than this; and I deſign to ſee that ſurprizing Appearance, if it pleaſe God I live ſo long: But methinks 'tis a little ungratefully done of the Moon and theſe lower Planets, ſaid ſhe, thus to Eclipſe him, or deprive him of any of his Light, when they receive all theirs from him; tho' I'm almoſt afraid our Earth doth ſo too; for ſince ſhe is a Moon to the Moon, it muſt often be interpoſed between the Sun and Moon, and therefore for a Time deprive the latter of the Light of the former.

UPON my word, Madam, ſaid I, you begin to run great Lengths, and go deep into the very Heart of Aſtronomy: And if you will pleaſe to read Dr. *Gregory*'s *Comparative Aſtronomy*, in the Place I before recommended to you, you will be glad

glad to see how rightly you have reasoned. Shall we proceed next, Madam, to talk about what they call the *Superior Planets*; and in particular about *Mars*, who next occurs in Order?

Mars.

YES, said she, we must take him in his Way; but I hope you *Astronomers* han't such terrible shocking things to say of him as the *Poets* have. Mr. *Dryden*, I remember, gives such a Description of *Him* and his *Temple*, as when I read it, chill'd me with Horror; and what is worse still, after he had enumerated all manner of Slaughters, Famines, Plagues, *&c.* he adds this:

These and a thousand more the Fane adorn,
Their Fates were written e'er the Men were born:
All copied from the Heavens, and ruling Force
Of this Red Star, *in his revolving Course:*
The Form of Mars, *high on a Chariot stood,*
All sheath'd in Arms, and gruffly lookt the God.

No, Madam, said I, we give him no such Power of doing Mischief in our *Hypotheses*; but make him as calm and as gentle as any of the Planets.

VERY well, said she, then begin, and say what you please of him.

THEY

THEY account the Diameter of *Mars*, Madam, ſaid I, to be about 4400 Miles, and therefore he muſt be much leſs than our Earth: And his Diſtance from the Sun is about 123,000,000 Miles; he revolves about the Sun in 687 Days nearly, and runs at the rate of 45000 Miles in an Hour.

WELL! ſaid ſhe, that is pretty good marching too, for a Man in Armour. Sir, pray go on.

MADAM, ſaid I, by ſome Spots which have appeared in him, the Time of his Diurnal Revolution, is by Mr. *Huygens* ſettled exactly at 24 Hours 40 Minutes; and the Motion of thoſe Spots hath alſo diſcovered that this Axis hath very little or no Inclination to the Plane of his Orbit; and therefore the *Martial* Inhabitants will have no ſenſible difference between Summer and Winter. *Huygens* thinks that the Colour of the Earth in him is blacker than that of *Jupiter*, or the Moon. His Light and Heat is twice, and ſometimes thrice, as weak as what we receive from the Sun. When he is in his *Quadratures*, as they call it, that is in the middle between his Conjunction with, or Oppoſition

tion to the Sun, he appears a little gibbose, and to a good Glass almost *bissected*; but when at Full, perfectly round and distinct. The Telescope hath not yet been able to distinguish any *Moon*, or *Satellites* moving round him; but that will not be a demonstrative Reason that there are none at all: for as they are at a great Distance from us, so they may be but small, and reflect but a weak and small Light, and therefore may not be visible. The Proportion of Heat and Light in this Planet, in comparison of ours, is not much above half.

O! said she, for all he looks *so red*, then I perceive the Planet is not so fiery as the Poets feign the *God of War* to have been. Pray, Sir, let us go on to *Jupiter*.

Jupiter. *See Fig.* IV.

THIS Madam, said I, is the largest of all the Planets, and you see by the general Scheme that he is much more remote from the Sun, than any of the Inferior Planets we have already been discoursing of, and therefore Heaven hath granted him a Supply of Light, by 4 Moons or Satellites, which revolve round him as our Moon doth round us; and these Moons, like the Satellites of *Saturn*, are so much less than their primary Planets, that

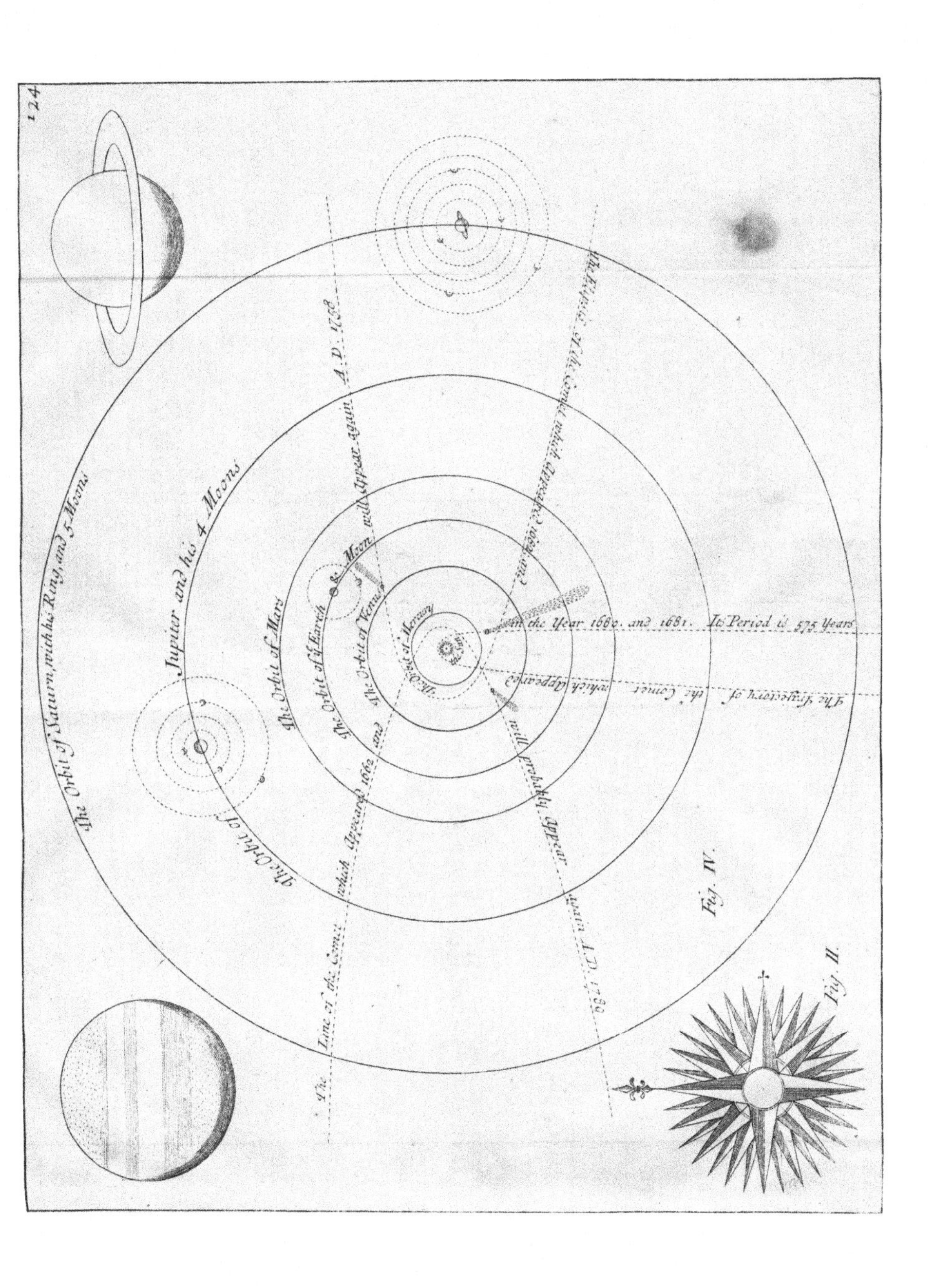

The material originally positioned here is too large for reproduction in this reissue. A PDF can be downloaded from the web address given on page iv of this book, by clicking on 'Resources Available'.

that they are not visible without long Glasses, and therefore were perfectly unknown until the last Age.

These secondary Planets suffer 4 kinds of Eclipses. (1.) When they are within the Shadows of their Principal. (2.) When the primary Planet is between them and us. (3.) When they are between their Primary one and us; for then 'tis difficult to distinguish of 2 Luminous Points one from the other. (4.) When they interpose between one another and our Eye, so as to hide one another from our Sight; which indeed happens but very rarely. And all these Attendants or Satellites, as well as *Saturn*'s, like the Moon, the Earth's most obsequious humble Servant, do always turn their Faces towards their Lords the primary Planets, about whom they revolve, and on whom they wait.

THIS, said she, exhibits a good Image of respect and regard in Servants and Attendants; I wish our Earthy ones would imitate the Celestial.

MADAM, said I, the Times of the Periodical Revolutions of *Jupiter*'s Moons round about him, are as follows:

The Innermost moves round him in 1 Day and 18 Hours, the second in 3 Days

Days 13 Hours, the third in 7 Days and almost 4 Hours, and the outermost in 16 Days and 16 ½ Hours. In the *Lexic. Technicum*, you will find a good deal more about this Planet, and how the *Eclipses* of his *Satellites* are calculated; and thence arises an easy way to find the Longitude on Shore: But I doubt it is not practicable at Sea. If this Evening happen to be clear, as (it promises well) I will shew you the Planet with his Attendants about him.

I SHALL long to see that Sight, said she; but pray go on, and tell me more about this noble Planet.

HIS Diameter, Madam, said I, is above 80,000 Miles; and the Quantity of Matter in him is about 220 times greater than that of our Earth; and his Distance from the Sun about 424 millions of Miles: He revolves round his own Axis in 9 Hours and 56 Minutes, and about the Sun in 11 Years and 10 Months: And so large is his Orbit, that he moves after the rate of about 24,000 Miles in an Hour.

THIS Planet, said she, makes a great Figure by the largeness of his Bulk, and the grandure of his Attendants; but pray what

what kind of Temperament hath the Air of *Jupiter*? I doubt it muſt be much colder than ours, and then I ſhall never deſire to be a *Jovian*.

MADAM, ſaid I, the Heat and Light of the Sun can't be above a 27th Part of what we enjoy here, and therefore it muſt be very dark, diſmal, and cold living there; and the weight of all Bodies will be double to what they are on our Earth.

NAY! ſaid ſhe, if the People be twice as heavy, and almoſt 30 times as cold as we are, even let them live by themſelves for me, I'll never hanker after going thither, but content myſelf with ſome *Jovial* Friends here in our dirty Planet, as Dr. *Burnet* called it; but I ſhall never have much value for his Judgment any more, that repreſented *Jupiter* as the Pattern of the fine *Antedeluvian* World. But pray, Sir, What Diſtance may theſe 4 Moons of *Jupiter* be from his Body?

THE neareſt, Madam, ſaid I, is about 130,000 Miles from that Planet; the ſecond about 364,000, the third 580,000, and the fourth or outermoſt is about a million of Miles diſtant from him.

WHAT

WHAT a fine Appearance must these 4 Moons make, said she, and what frequent Eclipses of the Sun, and of one another, do they produce! And if *Jupiter* hath any Ocean, and it ben't always frozen up like the *Baltick* in a hard Winter, what whisking Tides must they produce *there*, since our own Moon hath so great an Effect *here*, in that respect!

MADAM, said I, I see you don't only advance in *Astronomical*, but even in *Physical* or *Natural* Knowledge: that Speculation about the *Tides* of *Jupiter* is curious and new, and will be worth a further pursuit. But if you please we will now go on with our Planet's *Phænomena*. You see by the *Figure of Jupiter*, that besides a famous Spot by which his Diurnal Motion was determined, there are appearances in him like Swathes or *Belts*, as they call them: These they take to be moveable, and to be formed by the Clouds of this Planet, which seem, like our Trade Winds, to lie in Tracts parallel to the *Equator* of *Jupiter*.

AND if these are really Clouds, says she, won't it be a proof of *Jupiter*'s having

having a vaporous Atmoſphere about him, like that of our Earth.

It will doubtleſs ſhew, ſaid I, Madam, that he hath ſomething round him like our Air: but its Texture, Gravity, and Elaſticity may notwithſtanding be vaſtly different from that of Ours; but if by it you mean to inſinuate *that he is inhabited*, I entirely agree with you; for I take it, that ſuch an *Apparatus* as the making four Moons to revolve about, and to enlighten him; (as five ſuch there are alſo moving round *Saturn*, beſides his Ring) I take this, I ſay, to be a Demonſtrative Proof of both theſe Planets having *ſome kind of Inhabitants*, who have Eyes to ſtand in need of *Light*, as well as *other Senſes* proper for their Natures: For we never find Nature doing any thing in vain, but ordering all things with the moſt conſummate Wiſdom; and we muſt never believe ſhe would form Moons, where there are no People to be lighted by them.

Do you think, ſaid ſhe, that our Earth can be ſeen by the *Jovial* Inhabitants?

No Madam, ſaid I, by no means.

LORD! what vain Creatures we are, ſaid ſhe, in this Earthly Planet? What a buſtle do we make to extend our Power and Empire over it? But I'm mightily glad the impertinent and deſtructive Ambition of an *Alexander* or a *Louis le Grand*, can't be heard of in *Jupiter*; and I hope the Heroes *there* are always exerting themſelves for the good of their People. How vain is it alſo in ſome of our Divines, to ſuppoſe *Jupiter*, as well as the reſt of the Heavenly Bodies, to be made *only* for the uſe of Mankind? When yet, neither in him nor in *Saturn*, can the Place of our Habitation be ſeen. But pray, Sir, go on.

THIS, Madam, ſaid I, I think is all that is very remarkable about this famous Planet, except one thing more, which is indeed very conſiderable and ſurprizing: And that is this; that by the Eclipſes of *Jupiter*'s Satellites, made by the Interpoſition of his Body between them and our Eye, it hath been diſcovered that Light is in its Motion *Progreſſive*, and not *Inſtantaneous*, but that it takes up a determinate Time to come from *Jupiter* to our Eyes: For they have obſerved that theſe Eclipſes happen *ſooner* than they ought to do, by Calculation, when our Eye by the

An

Annual Motion of theEarth, *meets* theRays of Light reflected from them, whether at their *laſt going* out of the Sun's Light into *Jupiter*'s Shadow, or at their *firſt coming into that* Light afterwards; and theſe *Eclipſes* evercome *too ſlow* for the ſame Calculation, when we are going from thoſe Rays; and this is always in thatProportion, which implies that the Rays of Light go from the Sun to our Eyes in about 7 $\frac{1}{2}$ Minutes of Time: And on this Calculation it was, that what I told you before about the prodigious Velocity of the Rays of Light, was founded.

I SHALL look a little further, ſaid ſhe, into this Affair ſome other time; but pray let us now go on to talk about *Saturn*.

Saturn.

THAT outermoſt Planet in our Syſtem, Madam, ſaid I, is at a very great Diſtance from the Sun, about 777 millions of Miles; and the Time of his Revolution round him, is about 30 Years, or more exactly ſpeaking, in 10759 Days, 6 Hours, and 36 Minutes: And yet ſo very large is his Orbit, that he moves at the rate of about 18000 Miles an Hour; his Diameter is about 61000 Miles; and with regard to the Quantity of Matter in him, 'tis about 94 times as great as that of our Earth; but his Density is not much above a 7th part of that of the Matter of our Planet. And as to Light and Heat, 'tis probable that he hath not above a 90th part of what we enjoy by the Sun. Indeed in order to supply this great Defect of the Sun's Light, occasion'd by so great a Distance, our All-wise Creator hath furnish'd him with *Five Moons* or Attendants; the largest of all which, and which is the only one that is commonly seen, is the 4th in order from his Body; and he bears the name of the *Hugenian Satellite*, because first discover'd by Mr. *Huygens*. These Satellites of *Saturn* revolve round him in the *Plane* of his Ring (of which Ring I shall speak presently) and so their Circles make the same Angle with his Orbit, that the Plane of his Ring doth, which is about 31 Degrees. But the Orbit of *Saturn* himself is nearly coincident with the Plane of our Earth's Ecliptick, as are indeed the Orbits of all the primary Planets. It doth not yet, I think, appear, that *Saturn* hath any Diurnal Revolution round his own *Axis*; the Time of his Periodick Motion round the Sun, I gave you before; and those of his Satellites are as follows: The Innermost of these Moons revolves round *Saturn* in one Day,

Day, 21 Hours and 20 Minutes, and is distant from him about 146,000 Miles. The second is distant from him about 187,000 Miles, and performs his Revolution in 2 Days, 17 Hours and 40 Minutes. The third's Revolution takes up 4 Days, 13 Hours and 45 Minutes, and he is distant from the Centre of *Saturn* about 263,000 Miles. The *Hugenian* Satellite is about 600,000 Miles from him, and moves round him in 15 Days, 22 Hours and 40 Minutes. The last is 1,800,000 Miles distant from *Saturn*, and takes up 79 Days, 22 Hours in revolving round him.

'Tis highly probable that there may be more Satellites than these five moving round this remote Planet; but their Distance is so great, and their Light may be so obscure, as that they have hitherto escaped our Eyes, and perhaps may continue to do so for ever; for I don't think that our Telescopes will be much farther improved.

But the most surprizing and unparallel'd *Phænomenon* of all, in this Planet, is that which we call *his Ring*; which appears nearly as the Figure represents it, in an ordinary Telescope: 'Tis a vast Body of Earth, as is most probable, of

Vid. Fig. of Saturn.

 perhaps

perhaps 7 or 800 miles in Thickness, which at the Distance of about 21000 miles from *Saturn*'s Body, and with just as great a Breadth, is placed in a circular Arch, round about the Planet, in Figure much like the great Wooden Crane-Wheels, in which Men or Horses walk, to raise Goods, or to draw Water. 'Tis placed exactly over the Equator of *Saturn*, and is not any way contiguous to his Body, nor supported by any thing. The *Surface* of this Ring is not rough and full of Hills and Protuberances, as *that* of the Moon in most places is; but even and plain, as it is in those Regions of the Moon, which some, because of their great Evenness, have judged to be *Seas*.

THE Thickness of the Ring, comes not into Astronomical Observation, appearing but as a Line. And tho' the two broad Surfaces of the Ring reflect a good deal of strong Light, yet the marginal Surface of it, or its Edge or middle Part between the two eminent Surfaces, reflects hardly any at all. The Plane of the Ring is inclined to that of the Ecliptic, with an Angle of about 31 Degrees; and this Inclination in the Course of one entire Revolution of *Saturn* round the *Sun*, hath some Variation; being twice

greatest,

greateſt, and twice the *leaſt* of all. And this occaſions the Planet ſometimes to appear without any Ring at all, as when the *Sun* happens to be in the Plane of the Ring; and at other times, with *Anſæ*, as they call them, or *Handles* only; when but little of the Surface of the Ring can be ſeen: And at all other times the Ring will appear in an Oval Form, which ſometimes will be more, ſometimes leſs *oblong*.

I ſuppoſe, ſaid the Lady, it is at that critical Time when the *Anſæ* only appear, that *Saturn* puts on the Figure which *Hudibras* makes *Sydrophel* give him, that is, that its like a *Tobacco-Stopper*.

THAT is but a mean Ridicule, ſaid I, Madam; but I perceive it hath ſome Uſe; for it impreſſes itſelf, and the Thing, ſtronger on the Memory, than perhaps a more juſt and ſerious Deſcription would have done. But your Ladyſhip will ſoon be above theſe little Helps: And you will receive a great deal of Pleaſure, Madam, by reading what Dr. *Gregory* hath written about this Ring, in his Diſcourſe of *Saturn*, and in his *Comparative Aſtronomy*, ſo often recommended to you; where the moſt conſiderable Phænomena of this *Ring*, and of the *Satellites*, as they appear to an Eye ſuppoſed to be placed in *Saturn*, are explained and accounted for; or you may conſult the *Lexicon Technicum.*

I will, ſaid ſhe, attach myſelf heartily to that Book, as ſoon as I can: And after we have view'd this Planet with our Teleſcope, which I will ſit up any time of the Night to do, if you can afford me your Aſſiſtance. For theſe two ſuperior Planets have ſo many Wonders attending them, that I grow ſeriouſly amazed; and long to underſtand a little more of them, and to contemplate theſe wonderful Works of our great Creator. And indeed what a vaſt Field of Thought, what a new World of Speculation, do theſe new Diſcoveries open to us! How empty and ſtarv'd is a Mind unfurniſh'd with ſuch glorious Ideas!

What a rich Fund of Images is treaſured up *here* to embelliſh our Poetry? And yet I don't remember to have met with many Alluſions taken from theſe Things, except in a late Copy of Verſes preſented to her Grace the Dutcheſs of *Bolton*, where after the Poet had ſaid a great many fine and juſt things of of her, I now remember these Lines; the Beauty and Propriety of which, did not at first strike me so much as they do now, since I have been conversant with these Speculations.

BOLTON's *the Centre of Respect and Love:*
Round her like Planets, we at Distance move:
From her receive our Light, derive our Heat,
And still to'ards Her we tend and gravitate,
Just in Proportion to our Sense *and* Weight.

But now, Sir, said she, if you please we will leave off, *unbend*, and go to our Tea.

THE Lady plied her Telescopes, and pursued her Astronomical Studies with great Application and Success; and after some time, when I had the Honour to wait upon her again, she took me out into the Summer-house in the Garden, and then began thus with me.

SIR, said she, you have already taken a great deal of Pains to gratify a Woman's Curiosity; but I must beg you to indulge me yet a little farther, and to afford me a Lecture upon another Point; about which, as I am ashamed to trouble you

you, so I should be afraid to ask you, but that you have been so kind already, as to help me to get rid of many Fears and Terrors, too incident to our Sex: And if you can ease my Mind of this remaining Dread, I shall think you can do me a signal piece of Service.

You must know I have been tumbling over those Books of Astronomy, which you have bid me read; and tho' there be very many Things that I don't understand fully at present, yet there are some also that I know enough of, to be put into the Vapours by them.

The *Affair of Comets*, Sir, with their grisly *Beards* and horrid *Tails*, fright me almost out of my Wits: For god-sake therefore, tell me, as plainly as you can, whether my Dread is well grounded; Do they really *forebode* all manner of Mischief to Mankind, as well as *do* a great deal, when they come among us? What are they? Are their Motions natural, and accountable by Mathematical Calculation, as those of the Planets? Or are they miraculously sent hither as the Messengers of God's Wrath, and as the Executioners of his Judgments upon sinful Mankind?

MADAM,

MADAM, ſaid I, as to their *Preſages*, I take them to be entirely groundleſs; but they may be made (as almoſt any other of the Heavenly Bodies may, if God pleaſes) to become the Inſtruments of Evil and Deſtruction to any of the other Planets: but indeed it doth not plainly appear, ſince their Motions and Appearances have been of late more fully enquired into, that they *have any ſuch deſtructive Uſe*, or that they have actually done any real Miſchief in the Planetary World. There have indeed been ſome ſuch *Conjectures*; but as I take them to be *no more*, I will not trouble you with them now; becauſe I believe they will occur to you in your future Purſuit of theſe Studies.

I'M glad to hear you ſay ſo, ſaid ſhe, and I begin a little to be comforted: But pray go on, and compleat my Cure; for I don't care to be *drown'd* or *burnt* up by one of theſe extravagant Ramblers a Comet, before I am aware.

O MADAM, ſaid I, I perceive where you have been dipping; I will therefore give you the moſt ſatisfactory Account I can.

THE

THE Ancients, you must know, generally believ'd *Comets* to be only *Meteors*, like our *Firedrakes, &c.* and that they were no higher than our Regions of the Air; while some modern Writers placed them among the fixed Stars. But subsequent Observations, with good Instruments, and the Application of the Laws of Motion and Geometry, to Astronomical Enquiries, have now satisfied us almost to a Demonstration, that they are a *kind of Planets* revolving in determinate Periods round the Sun: But indeed the Orbits of many of them are so very oblong, *excentrick* or *oval*, as well as large and extended, that they can appear to us but very seldom; and when they do become visible, they exhibit Appearances which are very surprizing; for the lower ends of their Orbits are so very near the Sun, that when they come down into that part, or into their *Perihelion*, as 'tis call'd, they are actually heated and set on Fire by him to such a Degree, as not to get off again, without such dreadful *Beards and Tails*, as would really fright such as don't understand and consider how they come by them.

BLESS

Bless me! said she, why then if our Earth moved in such an Orbit, I see we might be easily destroyed and burnt up, by that very Sun, who now gives us cheering Light and kindly Heat!

'Tis very true, Madam, said I; for that great Comet which appeared here in the Year 1680, (and which I saw, and very well remember, tho' then but a Boy) went so near to the Sun, as to acquire a Degree of Heat above 2000 times as great as that of red-hot Iron: And if its Body was about the Size of our Earth, as it was judged to be, it won't be cool again this Million of Years: And yet it pleased God, that that Comet went away from us, without doing us any sensible Harm, that I know of; and so little do I fear being hurt by any of them, that I could almost wish another would appear, to help us to compleat the Theory of their Motions.

Nay, said she, if you that know so much of them are not afraid of them, I'm sure I won't be so for the future: Pray therefore, Sir, proceed and tell me what you can of the Number, Motions and Appearances of these Comets, how their

their *Beards* and *Tails* are formed; and how you account for the most eminent of their Appearances.

MADAM, said I, there have within this last 400 Years appeared to this part of the World but 24 Comets, (how much greater a Number there may be God knows, and perhaps subsequent Observations may discover more.) And of these according to the Observations of Dr. *Halley* and other Astronomers, *three* of them have had their Orbits, and Appearances so *very like*, and the Times of their appearing so very *equal*, that they have judged it very probable that those 3 Comets which successively appeared as three were in reality but one or the *same Comet* appearing at three several Times.

And the like they are inclined to judge of two others; that they also are but one, appearing at two different Times.

That great Comet that appeared here in the Years 1680, and 1681, was seen before in our Hemisphere, *A. D.* 1106; once before, about the Year 532; and also 44 Years before our Saviour's Birth: and therefore they conclude the Time of its Periodick Revolution round the Sun to 575 Years.

THE

The Time of the Revolution of *another Comet*, which they judge will appear again *A. D.* 1758, is 75 Years: *Another*, which probably may be seen here again, *A. D.* 1789, makes its *Ellipsis* round the Sun in 129 Years.

The Orbits of these three are described in Fig. IV.

WHAT Bigness do you take these Comets to have been of Sir, said the Lady.

MADAM, said I, they are generally of the size of the rest of the Planets, and have Atmospheres about them like our Earth: But then as all our Planets move pretty nearly in the *Plane of the Earth's Ecliptick*, these *Comets* are tied to no such Rules; for the *Planes of their Orbits* have very different, nay, almost all manner of Directions and Positions, and their Motions are all manner of Ways; some from East to West, others from West to East, some from South to North, and others a quite contrary way, *&c.* And yet their Motion is equable enough, and shews us this great Point; that as there can be no such solid Orbs as was imagined in the *Ptolemaick System*; so there can be neither any such thing as a *Plenum*, and no such *subtile Matter* as the *Cartesians* have invented to solve their *Hypotheses*:

But

But we may fairly conclude, that all the vast Spaces both between and beyond the Planetary System, are filled with no Matter capable of making any considerable Resistance to their Motions, but rather are an immense Void, or Vacuity.

I think that is a very probable Conclusion, said she; for if there were any quantity of resisting Matter, it must always obstruct a little, and by degrees must make very sensible Alterations in the Planets Motions; which I don't find to have been in Fact discovered; but sure, Sir, these Comets must go off to vast Distances from the Sun?

YES, Madam, said I, and therefore they are still more unfit than any of the other Planets, to be inhabited by such kind of Beings, as those of human Race; for the middle Distance of the Great Comet that appeared in 1680, was more than 5000 millions of Miles from the Sun; as its greatest Distance was above twice as much; and yet its least Distance was not above a 20,000th part of its greatest: so that in its whole Revolution, it would be subject to such Extremities, as that its *greatest* Degree of Light and Heat to its *least*, were above 400 millions

to

to one. And yet notwithſtanding this immenſe Extenſion of its *Ecliptick Orbit*, the Great and Allwiſe Architect of the Univerſe hath probably ſo adjuſted the *Centrifugal* and the *Centripetal* Forces, that it doth not quite leave the Sun, tho' it go ſo far from him, but returns again towards him, and revolves round him in a determinate Period of Years. *None of the Orbits* of any of theſe Comets yet known, are in or near the Plane of the *Earth's Ecliptick*; and therefore in their *Aſcent* from the Sun, tho' heated never ſo much by him, yet they won't come near enough to our Earth to *burn* us, or affect us with any ſenſible Heat; and therefore, Madam, your Fears of being *burnt in your Bed* by a Comet, I hope will vaniſh for this time.

WELL, ſaid ſhe, and ſo they will; but I love to know the Reaſons of things as well as any a Man of you all. But pray, Sir, what are the *Heads*, *Beards*, and *Tails* of theſe Comets?

MADAM, ſaid I, the Bodies of Comets are probably in Subſtance like our Earth; fixt, ſolid, and compact: Their Tails are probably long and very thin trains of Smoak and Vapours, emited

 from

from the heated or enkindled Body, Head, or *Nucleus*, as ſome call it, after their *Perihelion*, or after their having been at their neareſtDiſtance to the Sun , for then it hath been obſerved, that the *Tails of all Comets* have appeared *largeſt* and *longeſt*. In the *Lexicon Technicum*, under the word *Comet*, you will find a great deal ſaid about the *Phænomena* of *Comets*, their *Beards*, *Tails*, *&c.* from Sir *Iſaac Newton*, and other Authors; and there you will likewiſe find Conjectures about their Uſe in the Planetary Syſtem.

SIR, ſaid ſhe, I ſhall have recourſe to thoſe Books with a great deal of Pleaſure, and will trouble you no farther now with my Enquiries: I ſee Company appearing, let us forget our Aſtronomy a while, and trifle with them as agreeably as we can.

ABOUT a Month after our laſt Conference, I waited on the Lady in *London*, who after the uſual Compliments, began thus with me.

THO' you might be juſtly afraid to meet ſuch a queſtionary Creature as I am, I will own, I'm glad to ſee you in this Place; for I have a great many things to enquire of you, with relation to our late Conferences

Conferences in the Country. Ever since that I have been tumbling over Astronomical Books with the utmost Application; I have dipt a little also into the *New Physicks*, and I have been running over *your* Geometry, *your* Trigonometry, and *your* Spherick Projection, in order to use myself to Figures, and to get clearer Ideas of what the Astronomical Writers say: And tho' I believe I should have been frighted and deterred from beginning with *Geometry*, and the *abstracted Mathematicks*, yet I now find them so necessary that I am resolved to try at them, and will beg your help, when your Leisure will permit. But in the mean time pray tell me, Don't you think that the *Elementary Mathematicks*, and the *Newtonian Physicks*, or *Natural Philosophy*, might be taught to Gentlemen, or even to our Sex, in the easy and delightful way you have instructed me in Astronomy?

DOUBTLESS, said I, Madam, there is no one *really* Master of any Science, but he can communicate it to another in plain and easy Words, and render it intelligible to any common Capacity and inquisitive Genius.

WHY then, said she, if I have any Power or Influence over you, which sometimes you compliment me with believing, I would desire you by all means to attempt that, as your Leisure will occasionally permit you, and in the Intervals between your severer Studies; for I really think it would be of the greatest Use and Advantage, not only to our Sex, but even to your own: And I'm satisfy'd too, that many of our young Gentlemen grow vicious chiefly because they are idle, and having been taught nothing to improve their Minds, can have no Notions of the Rapturous Pleasures of Science.

I ENTIRELY agree with you in your Notions, Madam, said I, and your Commands shall be my Delight as well as my Duty; in the mean time, can I serve your Ladyship in any thing now?

YOU are very obliging, said she, to anticipate your Trouble, but we will lose no time in Compliments: What I want at present is, to be instructed farther by some *Diagram or Figure*, how by the Earth's revolving round the Sun in her Annual Motion, together with *that* round her *Axis*, the different Seasons of the Year

Year, Length and Decrease of Day and Night, *&c.* are accounted for. Have you drawn me such a Scheme as you once promised me, for this purpose?

I HAVE, Madam, said I, and here it is; I took it chiefly from Mr. *Flamstead*'s *Doctrine of the Sphere*; a Book, I dare say your Ladyship will one time or other dip into.

I HAVE seen it, said she, in Sir *Jonas Moor*'s *Mathematicks*, and perhaps may consider it further; for tho' I never design to attempt the Calculation or Construction of Eclipses, yet I shall be glad to know how the Astronomers do it. But pray, Sir, go on, and explain the Figures to me.

An Explication of Fig. V.

LET the Circle A B C D represent the *Earth's* Annual Orbit round the Sun, whose Centre is supposed to describe that Periphery, as it moves round the Sun from A towards B, in the Natural Order of the Signs, and from *Aries* to *Taurus*, *&c.*

THE Line ♈, ☉, ♎, represents the *Equinoctial Colure*, and the other ♋, ☉, ♑ standing at right Angle to it, is the *Solstitial Colure.*

N.B. The Figures ♈, ♋, ♎, ♑, should be placed in the Circular Line A B C D.

The 4 lesser Circles *d o t i* represent the Earth's quadruple Position in the 4 *Cardinal Points*, as they call them, *i. e.* at the 2 Equinoxes, and the 2 Solstices, and the Line *d t* at right Angles to the Colures, may fitly enough be called the *Horizon of the Earth's Disk*, because it separates that half part of the Earth which the Sun shines on, from the other which lies behind in the Dark.

BUT pray, said she, what do you mean by the Earth's *Disk?*

I USE the Word, Madam, said I, because you will frequently meet with it in your reading; it signifies that round appearance of the Sun, Moon, or Earth, which we suppose to be the Object of any Spectator's View; and therefore the Earth's Disk is the appearance of that half of it, which because it is enlightened by the Sun, is seen by any remote beholder.

VERY

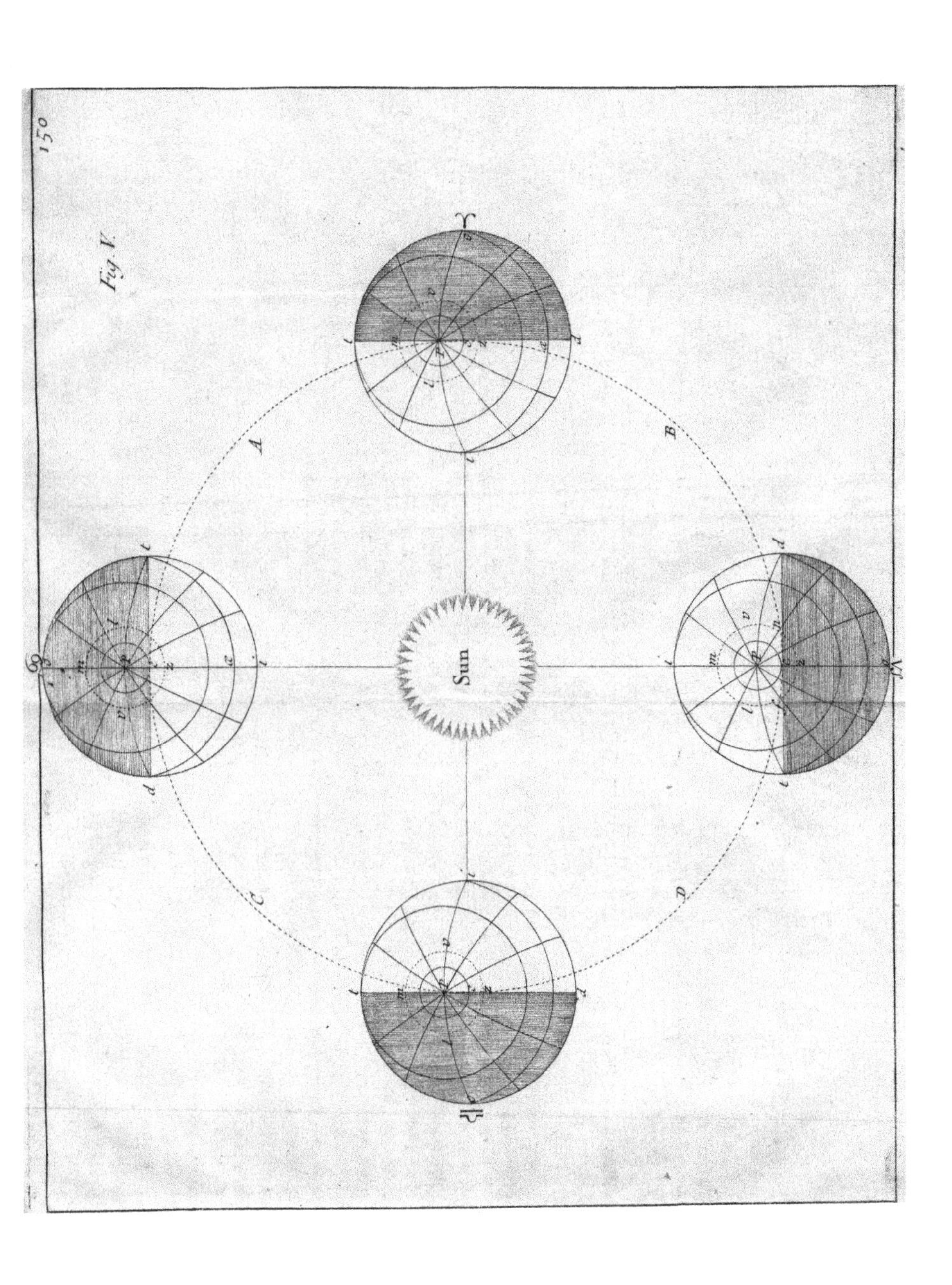

The material originally positioned here is too large for reproduction in this reissue. A PDF can be downloaded from the web address given on page iv of this book, by clicking on 'Resources Available'.

VERY well, Sir, said she, pray go on.

IN these 4 Figures of the Earth, the Spectator's Eye is supposed to be below under the Earth's Centre *e*, which Centre always moves in the Circle A B C D. To an Eye so placed, the Circle *d o t i*, which divides the Earth's Upper Hemisphere from the Lower, will appear to lie in, or be *coincident* with the *Plane* of the *Eclipt[i]ck*; and therefore that may be called the *Ecliptick on the Earth's Globe.*

THE *North Pole* of the Earth, or the upper End of the *Axis*, about which her Diurnal Motion is made, will then appear to be at *P*, 23° 30′ distant from *e* to the Pole of the Ecliptick; and if you draw a Line thro' those Points connecting the two Poles, that may be called the Line of *Direction of the Earth's Axis*; and if produced, it will be coincident with, or parallel to the great *Solstical Colure*, and therefore will describe *such a Line* on the Earth, to which, when the Sun's Rays run parallel, or whenever the Earth's Centre is in the Points ♑ or ♋, then will the *longest*, in the latter, and the *shortest Days*, in the former Case, happen to all the Inhabitants of the Earth.

This Line of Direction *P e*, is always found parallel to the Line ♋, ☉, ♑, during the whole annual Revolution of the Earth.

PRAY, said she, what occasions this *Parallelism of the Earth's Axis?* which I have read much of.

MADAM, said I, 'tis not any *new Motion*, superinduced into the Earth, but only her keeping to the first Position or Direction of that Diameter about which she revolves; which she must always do, without it be changd by the Will of the Great Creator, who at first appointed it to be so as it is, But if you please, I will go on.

A Line drawn perpendicular to the Earth's Axis, will represent on the Earth the *Equinoctial Colure*, and will always be parallel to the Great Equinoctial Colure ♈, ☉, ♎; and whenever the Sun Rays run parallel to this Line, which they will do, whenever the Earth is in ♈ or ♎, then will the *Days and Nights be equal* all the Earth over: For you see that as the Earth revolves round its Axis *t P e d*, all Circles described on the Earth, from the Pole *P*, *i. e.* such as are the *Equator* and all its Parallels, will be just one half in the

Fig. VI.

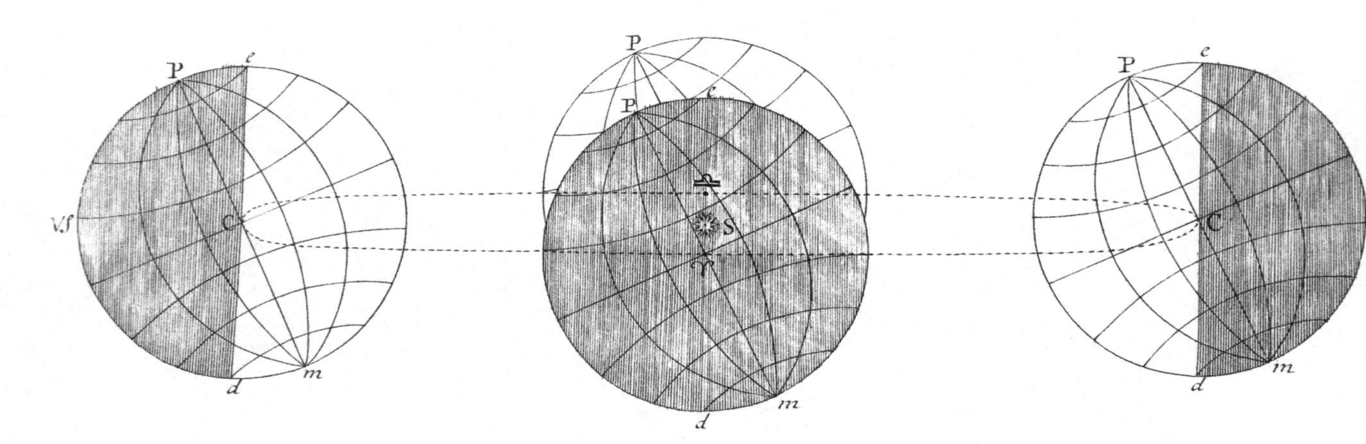

the Light, and the other half in the Dark.

The Angle made between the Earth's Axis and that of the Ecliptick, may be learnt best from *Fig.* VI. *Fig.* IV.

In which the Ecliptick Line ♈ C ♎ *e* represents the Earth's Annual Orbit, as view'd by the Eye, at a vast Distance, and when the Eye is placed a little above its Plane: Here let *e* be the Pole, and *e d* be the Axis of the Earth's Ecliptick, which you must suppose to be every where at right Angles to the Plane of the *Great Orbit*; and let *P* be the Earth's North Pole, *P m* the Earth's Axis, about which the Earth turns from West to East in 24 Hours; and suppose the Angle *P C e* to be always the same, *viz.* $23^{\circ} 30'$.

These things being supposed, it will be plain that every Point on the Earth's Surface, will, as the Earth revolves in her Diurnal Motion, describe a Circle about the next Pole: And when you consider, that every *such Point* is *Vertical* to the Earth's Centre, and answering to what hath usually been called the *Zenith*, or *Vertex*, in the *Ptolemaick* Projections, the Circle so described, is very properly called the *Path of the Vertex*, because 'tis a Track or Line made by the Motion of that Point.

I FANCY I ſhall conceive this right, ſaid ſhe, when I get to my Globe, for then if I bring *London* into the *Zeinth*, the Point on the Globe repreſenting *London*, is, I ſuppoſe, what you call the *Vertex*; and if I turn the Globe round its Axis, I ſee that Point will deſcribe a Circle, parallel to the Equator; and ſuch a Parallel, I take it, you call the *Path of London*.

EXACTLY right, ſaid I, Madam, and no one could have explained it better. I think then, we ſhall now go on with Pleaſure.

In ſuch Projections as theſe 4 Figures of the Earth in *Fig.* V. a Circle equally diſtant from both the Poles, muſt be the Earth's *Equator*; and the Diſtance of any Place from that Circle, will be the *Latitude* of that Place; and therefore half the Diameter of any *Path* will be the Sine Complement of the *Latitude* of any Place, deſcribing that Path.

If you take any Place on the Earth, and make a Circle paſs thro' it, and the two Poles, that will be the *Meridian* of that Place.

That Point in the Earth's Periphery, which is oppoſite to the Sun, or which is found

found by a right Line drawn from the Earth to the Sun, is called the *Sun's Place in the Ecliptick.*

i P o, and *t P d* in Fig. V. represent the Earth's first Meridian, in each Pair of the opposite Circles.

m v z l, represents the Circle made by the *Vertex* of *London*; as that within doth the *Northern Polar Circle*; and the next without it, the *Northern Tropick.*

By the Figure it will appear plain, that since the Sun enlightens but one half of the Earth's Globe at a Time, if the Earth be in ♎ or ♈, the *Horizon of the Disk* will then coincide with the *Solstitial Colure*; and therefore as the Earth turns round her *Axis*, which now is coincident with the Line *d t*, the Paths of the *Vertices*, or the Equator and all its Parallels will be bissected by the Line *d t:* and while any particular place on the Earth, or any *Vertex* is in the Light Part *t i d*, the Inhabitants of it will see the Sun; and therefore to them it will be *Day:* And while it is in the Dark Part, it will be *Night* to them.

But when the Earth is moved on, either from ♈ to ♋, or from ♎ to ♑, the *Line of Direction* will coincide with the *Solstitial Colure*, and the *Horizon of the Disk* will become

become at Right Angles to it on the Pole of the Ecliptick *e*. Wherefore, when the Earth is in ♑, all places between the two Poles of the Earth and the Ecliptick, and the *entire Artick Circle*, will, now you ſee, be illuminated in their whole Revolutions, as the Earth revolves round its Axis *i P o*. The *Vertexes* therefore will ſee the Sun, each one longer than 24 Hours, according as it is more or leſs diſtant from the Pole of the Globes; and thoſe that lie under the *Artick Circle*, touch the *Horizon of the Disk*; and conſequently at this time of the Year, *viz. June* 10, they will ſee the Sun 90 Degrees from the *Vertex*, both on the North and South of their Meridian; ſo that aſſoon as he is Set, he will immediately Riſe again; and conſequently they have no Night: But all *Paths* without this, you ſee, do *cut*, or get within the *Horizon of the Disk*; and ſo will have their Days longer than their Nighis, in proportion to the Quantity of the enlightned part of their Path, to the dark one; *i. e.* at *London*, As the Ark *n v m l f*, is to the Ark *n z f*; which is above Two to One: and therefore the Days will then be above 16 Hours long, and the Nights ſcarce 8.

Again, while the Earth moves from ♎ thro' ♑, and ſo on to ♈, you ſee the *North*

North Pole of the Earth is all that time in the Light part of the Disk; which ſhews you that to ſuch as live under that Pole there will be 6 Months Day. But while the Earth runs on from ♈ thro' ♋ to ♎, that Pole will, you ſee, be in the dark part of the Disk; which ſhews that then, under the Poles, there will be 6 Months Night. For indeed, when the Earth is in ♋, all things will be the very reverſe of what they are when ſhe is in ♑; *i. e.* the Nights longer than the Days, *&c.*

But when the Earth is in ♈ or ♎, the Axis of the Earth's Revolution being *d t*, (the *Horizon of the Disk*) juſt one half of the Equator, and all its Parallels will be enlightened, and the other half in the Dark; and therefore the Days and Nights muſt be equal all the World over.

SIR, ſaid the Lady, if you can part with this Figure, I will look it over more carefully another time, when I am by my ſelf. In the mean time I have another trouble to give you, if you will oblige me in it; and that is to get me a ſight of the famous *Orrery*, which I have heard you and others ſo often ſpeak of; and which I think was made by Mr. *Rowley*, the famous Mathematical Inſtrument-Maker,

Maker, and Master of the Mechanicks to the King; and whom I find you have always recommended in your Books, as the best Workman of his Profession.

I shall stay in Town about a Week longer, and will enlarge my Time a Day or two, rather than miss seeing so instructive and curious a Piece of Ingenuity.

MADAM, said I, the *fine Instrument* of that Name, which Mr. *Rowley* made for the *East-India* Company, is now luckily in a Place where I can come at it; I will go thither to morrow, and then appoint you a Day when I will wait on you to see it.

The

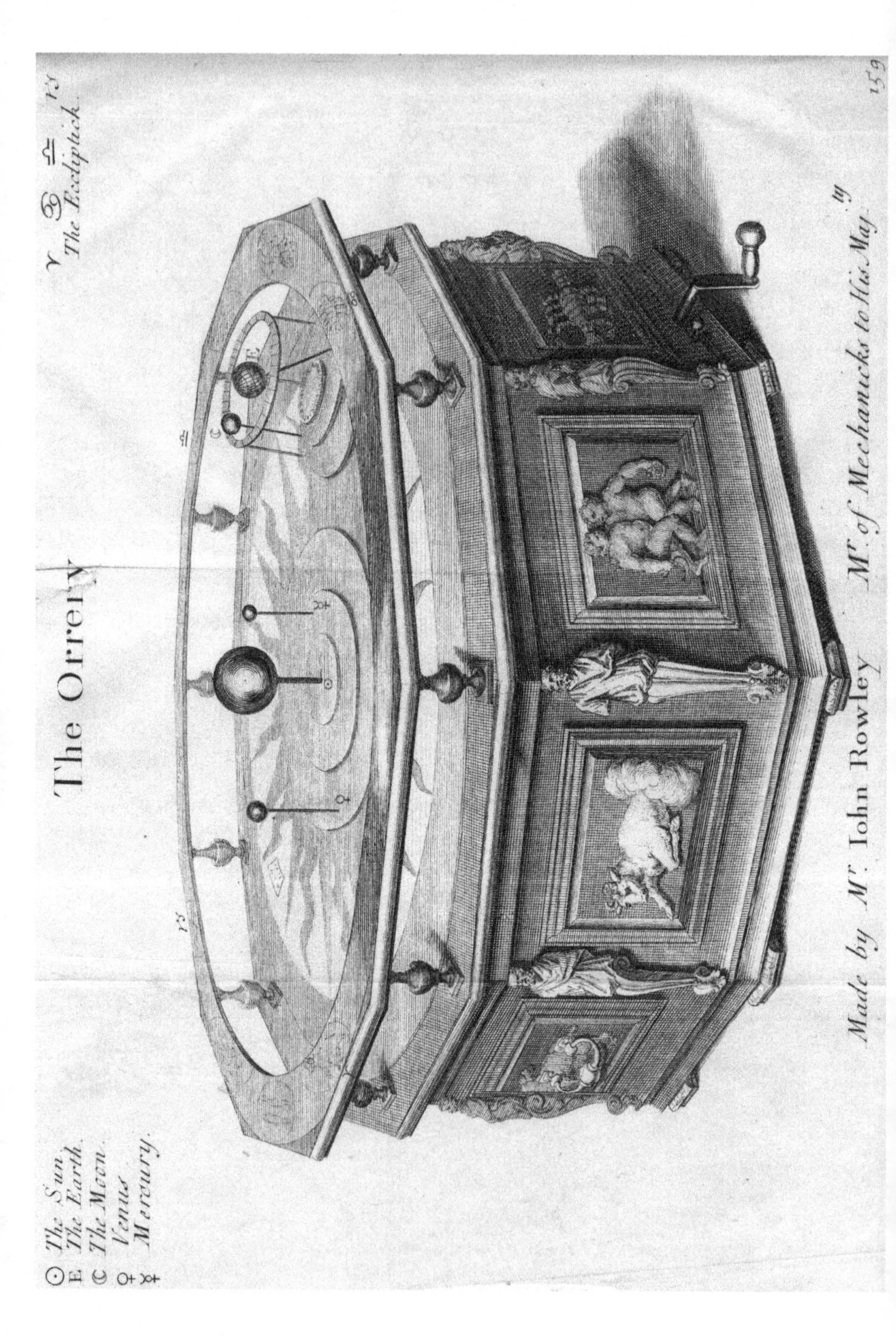

The material originally positioned here is too large for reproduction in this reissue. A PDF can be downloaded from the web address given on page iv of this book, by clicking on 'Resources Available'.

The Description of the Famous Instrument called the ORRERY; *made by* Mr. John Rowley, *Master of the Mechanicks to the King.*

WITHIN a Day or two, I obtained for the Lady a sight of the *Orrery*; she desired we might have no other Company but one young Lady more of her Acquaintance; because, said she, I shall ask so many Questions, as perhaps will shew my *Impertinence* to those who are not acquainted with Things of this Nature, and my *Ignorance* to those who are.

Assoon as the Instrument was taken out of its Case and set upon the Table, she expressed herself mightily pleased with the cleanness and cleverness of the Workmanship of it; for indeed the Outside of it is very rich and beautiful. The Frame is of fine Ebony richly adorned with twelve silver Pilasters, in the form of *Cariatides*; and with all the Signs of the Zodiack, cast of the same Metal, and placed between them; the Handles were also

alſo of Silver finely wrought, with the Joints as nice as ever were ſeen in the Hinges of any Snuff-Box: On the Top of the Frame, which was exactly circular like the Horizon of a Globe, is a *broad ſilver Ring*, on which the Figures of the 12 Signs are exactly engraved; with two Circles accurately divided; *one* ſhewing the Degrees of each Sign, and the *other* the Sun's Declination, againſt his Place in the Ecliptick, each Day at Noon.

Ecliptick and Zodi-ack.

The Nature and Uſe of theſe Circles the Lady perfectly underſtood, from what ſhe had before learned; and therefore in her pleaſant way, ſhe began thus:

IF ſo much Art and Expence be beſtowed upon the *Outſide* of this curious Machine, I don't doubt but the *Inſide* of it is at leaſt equally curious and uſeful: And therefore I muſt deſire you, Sir, ſaid ſhe, to begin quickly, and to ſhew it all to me, as the Man doth the Tombs at *Weſtminſter*; tho' I hope you won't be always in the ſame haſt, nor imitate his precipitant Manner, and awkward Tone of Speech; but do it ſlowly and diſtinctly, allowing me time to think and conſider about it, and to ask you all the Queſtions I have a mind to.

MADAM,

MADAM, ſaid I, you know, you can determine and command me, as you pleaſe.

This *Silver Plate* on which the Signs of the *Zodiack, &c.* are drawn, repreſents the Plane of the Great *Ecliptick* of the Heavens; or *that* of the Earth's Annual Orbit round the Sun; which as it paſſes thro' the Sun's Centre, ſo its Circumference is made by the Earth's Centre's Motion; and which for the better advantage of View and Sight, is here, you ſee, placed parallel to our Horizon.

The large *gilded Ball* which ſtands up, you ſee, here in the midde, not *upright*, but making with the Plane of the Ecliptick an Angle of about 82 Degrees, is ſo placed to repreſent the *Inclination of the Sun's Axis*; and which being pretty near the Centre of this Orbit, repreſents the Sun.

PRETTY *near*, ſaid ſhe; why is not the Sun then exactly in the Centre of that Circle which you call the Earth's Orbit?

NO, Madam, ſaid I, nor is that Orbit exactly a Circle; but an *Oval* or *Ellipſis*.

As in this Figure which I will now draw

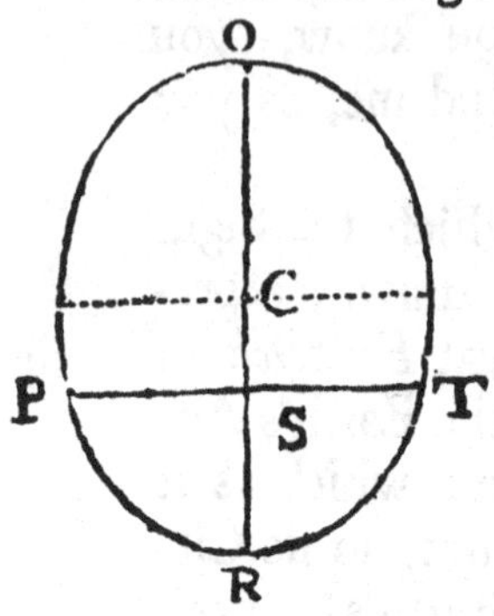

with my Pencil and ſhew you: Let the Curve Line *POTR* repreſent the *Orbit* of the Earth revolving round the Sun, which is placed not in *C* the Centre, but in *S*, a Point in the longer Diameter, which they call the *Focus*: The Diſtance between *C* and *S*, is what in the *Ptolemaick Syſtem*, was called the *Eccentricity*, and expreſſes how much the Earth's Orbit differs from being a True Circle. And the Contrivance of this Inſtrument is ſo admirable, that you will ſee by and by, when I ſet it a going, this *Eccentricity*, and that of the other Planets will be plainly ſhewn to your Eye, in the ſame proportion as they are in the Heavens.

PRAY, Sir, ſaid ſhe, go on; I find I ſhall come to underſtand this better, when I come again to read Dr. *Gregory*, and Mr. *Whiſton*.

MADAM, ſaid I, you ſee here two *little Balls* ſtanding upon two Wires, at different Diſtances from, but pretty near the

the Sun; the innermoſt of theſe is deſigned to repreſent *Mercury*, the other *Venus*.

BUT why are they placed, ſaid ſhe, upon thoſe two Wires; they ſtand perking up like the Traitor's Heads upon *Temple-Bar*; I hope Mr. *Rowley* hath not diſcovered that they have committed any late Treaſon againſt their *Sovereign the Sun*.

NO, no, Madam, ſaid I, they are very *Loyal Planets*; the Contrivance is only to bring their Centres to be, ſometimes in, and always pretty near the *Plane* of the great Ecliptick, (and by the by the Plane of their Orbit, always paſſes thro' the Sun, and interſects the Ecliptick in two Points, which they call *Nodes*) and this Poſition is contrived in order to ſhew us what Appearances they do really exhibit in their ſeveral Revolutions round the Sun. For the ſame Reaſon you ſee the *Earth* and *Moon* here placed likewiſe on Wires or Pins, that their Centres may get ſometimes actually into, and always be pretty near this Plane of the great Ecliptick; for ſo the Orbits of all the Planets are really placed in the Heavens.

I LIKE that *pretty Ivory Earth* very well, said she, as I do the *Golden Sun:* But pray why doth the Earth's Pin stand *inclining* so, and not *upright?*

MADAM, said I, that is to represent also the *Angle that the Earth's Axis*, or that of the *Equator*, makes with the Axis of the *Ecliptick*; which latter, in this Instrument, being perpendicular to the Horizon, the Earth's *Axis* is placed so as to make an Angle with the Plane of the Horizon of $66^{\circ}\frac{1}{2}$; or dipping down from the Zenith just $23^{\circ} 30'$, which you know is the Angle made by the two Planes of the Equator and Ecliptick. And as the Earth in each of her annual Revolutions round the Sun, always keeps her own Axis parallel to its self; so you will see, by and by, when the Instrument moves, that this *Terella*, or little *Ivory Earth*, will do so too, as it takes its Tour quite round the *Golden Sun* in this Instrument.

I LONG to see that, said the Lady, very much; but I suppose I must suspend my Inclinations, till you tell me 'tis fit they should be gratified.

MADAM,

MADAM, ſaid I, 'tis beſt to conſider the ſeveral Parts of the Inſtrument, firſt ſeparately or ſingly, and then the ſeveral Motions and *Phænomena* will appear in the better and more inſtructive Light: Therefore if you pleaſe, we will go on.

You obſerve Madam, ſaid I, *another Wire* here, ſtanding cloſe to this Silver Circle, and which hath a *Ball* upon it, whoſe Centre is in the Plane of that Circle: This is deſigned to repreſent the *Moon*; and the Silver Circle repreſents her Orbit round our Earth, the Plane of which always runs thro' the Earth's Centre, and the Figures that are engraved upon it, ſhew her Age, from one New Moon to another.

WELL! ſaid ſhe, this is mighty inſtructive! I long to ſee the Earth and the Moon move, but I know I muſt have patience: I ſuppoſe the Moon's Globe being *black* on one ſide, and *ſilvered white* on the other, is deſigned to repreſent her *Phaſes* as they call them, of which you have ſhewed me ſomething before.

'TIS ſo, Madam, ſaid I; and you will ſee this Machine ſo admirably contrived, that

See Fig. of the Orrery.

that what I told you of the Moon's monthly Revolution, will shew it self to be in fact true here; for the *Lunula* here, will turn round its own *Axis*, at the same time as it moves in this *Silver* Orbit round the *Terella.* And in reality, Madam, I can't blame your eagerness to see the Machine put into Motion, when I see how well you understand it, and know what it ought to do: And therefore you shall be detained no longer, than while I desire you to take Notice of this Hole in the great Brass Plate that covers all the Movement, and of this moveable *Index* here on the silver Ecliptick. You see there are on the former some Figures engraved; they are the common solar *Years:* and by taking the Instrument to pieces, it may be set to this present Time: And the Planets, by means of an *Ephemeris*, may be set to any particular Time also. So that if a Weight or a Spring, as in a Clock, were applied to the *Axis* of the Movement, so as to make it move round once in just 24 Hours, these *Representative Planets*, which you see here, would all perform their Motions round the Sun and one another, exactly in the same Order and regular Manner, as their *Originals* do in the Heavens; and this would then be

a true

a true *Celeſtial* or *Aſtronomical* Clock, which would ſhew the *Aſpects*, *Eclipſes*, and other Phænomena of the Sun and Planets, for ever. But becauſe this would be inſtructive only in that ſlow tedious way to ſuch as could have daily recourſe to it, Mr. *Rowley* hath contrived, by a Winch or Handle, to turn the Axis ſwiftly round about, and by that Means to ſhew all the *Phænomena* or Appearances in a very little Time, as you ſhall ſee I will now proceed to do; for by turning this Handle backward or forward, you may ſee what *Eclipſes*, *Tranſits*, *&c.* have happened in any Time paſt; or what will happen for any Time to come, without doing any injury to the Inſtrument.

I am amazed at the Thought and Contrivance of this Inſtrument, ſaid ſhe, and I doubt not ſhall receive a prodigious Pleaſure when I ſee it put into its proper Motions: But pray, Sir, let me firſt ask you, *Are all the Planets here?*

No, Madam, ſaid I, (for I ſee nothing can 'ſcape your Ladyſhip's diſcernment) here are only ſhewed the Orbits of *Mercury*, *Venus*, the *Earth*, and the *Moon*;

for the others are at too great a Distance to be brought into the Instrument, if any tolerable Proportion be observed between its Parts: And indeed, by what you will see of the Motion of these Three Planets, and of the Earth's *Satellite*, the *Moon*, you will easily know what the *Phænomena* of the Superior Planets and of the other *Satellites* would be, if they could be here shewn; as they cannot well be without embarrassing the Instrument with a vast Number of Wheels more: And it hath almost 100 already.

But now, Madam, I will fix on the Handle, and begin to put the Instrument in Motion.

One entire turn of the Handle answers to the *Diurnal Motion* of the Earth round its *Axis*, as you will see by the Motion of the *Hour Index*, which is placed at the foot of the Wire on which the *Terella* is fixed; and which you perceive moves once round as I now with my Hand turn the Spindle of the Machine round, after the same manner. You will take Notice also, that the Instrument is so excellently formed, that I can make the Motion tend either way, forward or backward; and turn it about after the same manner, 'till I bring the Earth to answer to any Degree

or

or Point of the *Ecliptick.* As for Inſtance, I will move it about till I bring the Earth to the firſt Point in *Aries.* Then you ſee, to an Eye placed on the Earth, the Sun will appear to be in the *Oppoſite Point*, that is, in the firſt of *Libra.*

BUT Sir, ſaid ſhe, I perceive as you turn the Earth about, the ſilver Circle on which the Moon's Age is placed, and which I think you ſaid repreſented her Orbit, *riſes and falls*; What is the meaning of that?

MADAM, ſaid I, you know the Moon's Orbit is not exactly in the Plane of the Ecliptick; but makes an Angle with it of between 4 or 5 Degrees: And juſt ſo much this Circle riſes above and ſinks below the great Ecliptick, according as the Moon hath North or South *Latitude*, and juſt as *much* as that *Latitude* is: And you will obſerve *two little Studds*, which are placed in two oppoſite Points of this *ſilver Circle*; they are deſigned to repreſent the *Moon's Nodes*, or the Points of Interſection of *her Orbit*, with *that* of the Ecliptick: Of which, more by and by.

O!

O! pray! move on, Sir, ſaid ſhe, this is amazingly fine: I fancy myſelf travelling along with that little Earth in its courſe round the gilded Sun, as I know I am in reality with *that* on which I ſtand, round the *real* one.

You ſee, Madam, ſaid I, that one entire turn of the Handle is, as I ſaid before, a Natural Day: Now, if you pleaſe to take off *one of the broadeſt of your Patches*, and make it a *Spot* upon the *Golden Sun* there, you ſhall ſee that your Patch will move quite round in 25 Days, or 25 turns of this Handle; and that will ſhew you how by the Motion of the *Spots* in the *real Sun* the Aſtronomers diſcover'd he had ſuch a Motion round his *Axis*, as you ſhall ſee Mr. *Rowley* hath given here to his *Repreſentative*.

Sun's Motion round his Axis.

Well, ſaid ſhe, ſince even my Patches muſt become Aſtronomical, I will ſtick one upon this *Fictitious* Sun; but I muſt own I don't love thoſe *Spots* upon the *Natural* one; nor to have any of his Face hid, or his Heat impaired: But ſhew me to what part of the Sun this Patch is to be preferred.

Spots.

Please

PLEASE to stick it, said I, Madam, just against the first Degree of *Aries*, and in the middle of the Sun's Body, —— Very well! Now you will see that as 365¼ of these turns of the Handle will carry the Earth quite round in the Ecliptick; so 88 will make *Mercury* perform his Revolution, and 244 Turns will make *Venus* move quite round the Sun.

Mercury.

Venus.

Twenty seven Turns and a little more than a quarter of one, you shall see, will carry the *Moon* round in her Orbit; in which time you will observe she always turns the same Hemisphere towards the Earth.

Moon's Periodic Month.

Take Notice also, Madam, that now I have just made 12 Turns and an half, which hath carried your Patch to the opposite part of the Sun.

AND shall I ever see it again, said she, shall I ever recover the *Solar Traveller*?

YES, Madam, said I, you may have it again; but pray keep it for hereafter only for such Uses; and don't replace it on your Face; for I am as angry at *Patches* in a *good Face*, as you are at *Spots* in the *Sun*; and for *your* Reason, because

becauſe I would not have any part of it hidden from me.

But the Handle goes on; a Turn or two more will have carried the *Moon* half round in her Orbit; obſerve how ſhe moves: 'Tis now 25 Turns, you ſee, your Patch is come ſafe about to you; off with it.

No, ſaid ſhe, there it ſhall ſtick till we have done, ſince you won't have it be on my Face any more: I love dearly to ſee it turn round; and perhaps ſhould I put it on, it may make my Head turn quite round too, as I think it begins to do already without it; but pray turn on your Handle however.

MADAM, ſaid I, at the end of 27 Turns and a Quarter, you ſee I have made the Moon perform her Revolution round the Earth: *Mercury* is got about a third part of his way; and in 17 Turns more will have finiſhed juſt half his Revolution. And *Venus*, you ſee, will then have advanced a fifth part of her Way, in proportion to the Magnitude of her Orbit: And the Earth alſo hath traverſed in the Ecliptick the Diſtance of above three Signs.

And

And by thus revolving the Earth and Planets round the Sun, you may bring the Inſtrument to exhibit *Mercury*, and ſometimes *Venus*, as directly *interpoſed* between the Earth and the Sun; and then they will appear as *Spots* in the Sun's Disk; as I hinted to you before, *p.* 114. And this Inſtrument ſhews alſo very clearly the Difference between what they call *Geocentrick* and *Heliocentrick* Aſpects, according as the Eye is placed in the Centre of the Earth or Sun.

WELL, ſaid ſhe, I have no Words to expreſs the Pleaſure and Satisfaction I receive from this moſt Curious Engine, nor the Amazement the wonderful Contrivance of it gives me. Were my Fortune but half as great as my Curioſity, I would have one of theſe Inſtruments aſſoon as poſſibly I could get it, and then without being beholding to any of you *He things*, I would turn it about myſelf, till I made it do all I had a mind to. And I wiſh now, that I could ſee the Inſide of it; and underſtand what Numbers of Teeth and Pinions he hath made uſe of, to produce theſe various Motions.

MADAM,

MADAM, ſaid I, that can't be done without the Hand of Mr. *Rowley* himſelf: But our moſt Excellent King having the ſame Deſire and Curioſity as your Ladyſhip, he took it all to pieces before his *Majeſty*, and to his great Satisfaction ſhewed him every Part of the Contrivance.

WELL, ſaid ſhe, ſince I can't have that Satisfaction now, pray proceed to let me know as much of it as you can.

Periodick and Synodick Months.

MADAM, ſaid I, you will next be pleaſed to ſee the Difference between the Moon's *Periodick* and *Synodick* Month, and the Reaſon of it, very plainly here ſhewn to the Eye: I have now turned the Handle round till I have ſhewn you juſt ſuch a Period, as the Time between *our firſt* New Moon, when the Earth was in the firſt Point of *Aries*, and the *preſent one:* and at the Earth's Place in the Ecliptick, where this happens, I will ſtick this bit of Paper; and turning 27 $\frac{1}{4}$ turns of the Handle more, you ſee, I have brought the Moon again to be exactly interpoſed between the Earth and the

the Sun; and then you know it will be *New Moon* to us; but you ſee the Line of the *Syzygy* is not right againſt the bit of Paper, but behind it; and it will require two Days time or *two Turns* more, before it will get thither.

I THINK the Reaſon of that, ſaid ſhe, appears here very plain; becauſe in this 27 Days the *Earth* advances ſo far forward in her annual Courſe, as is the quantity of the Difference in time between the Moon's two Months. But pray, Sir, ſaid ſhe, won't this naturally carry you to ſhew me how the Eclipſes are formed?

YES, Madam, ſaid I, and that is all which is material, that I have left to ſhew you.

You know, Madam, the Aſtronomical Books tell you there can be no Eclipſe of either Sun or Moon, but when the Moon is in or near the *Nodes:* And this will be here very plainly ſhewn to you by the means of this Thread, of which if you pleaſe to take that End, we will extend it ſo as to repreſent the Line of the *Syzygies:* I will turn the Handle about till the next Conjunction of the Moon comes to be in or near the *Node,*

or

or in the Plane of the *Ecliptick*; and then you shall see there will be an *Eclipse of the Sun.* You see I have turned the Handle about 27 times; but now the Centres of the Sun, Earth, and Moon are not near in a Right Line, as the Thread shews you; and therefore there will be no *Eclipse of the Sun:* But you see now at the (*a*) *Full Moon*, the Line connecting the three Centres, is very near the *Node*; therefore there will be an *Eclipse of the Moon:* And (*a*) now, you see, there is an *Eclipse of the Sun*; which is *Central*, when all the *three Centres* above mentioned come into this Thread thus stretched in the Plane of the *Ecliptick*; and *Total*, when the Moon is in her *Perigæum*, at the greatest Distance from the Sun, and nearest to us.

(a) *After I had turned it round several Times till it hapned so.*

But in order yet farther to shew the *Solar Eclipses*, and also the several Seasons of the Year, the Increase and Decrease of Day and Night; and the different Length of each in different Parts of our Earth, Mr. *Rowley* hath this further elegant Contrivance.

He hath provided this little Lamp to put on upon the Body of the Sun; which casting, you see, by the Means of a Convex Glass, and the Room made a little dark,

dark, a strong Light upon the Earth; will shew you at once all these things; first how *one half* of our Globe is always illuminated by the Sun, while the other *Hemisphere* is in the dark; and consequently, how Day and Night are formed, by the Revolution of the Earth round her *Axis*; for as she turns from *West* to *East*, she makes the Sun appear to move from *East* to *West*. And you will please to observe also, Madam, that as I turn the Instrument about in Order to shew you the several Seasons of the Year, and the Length and Decrease of Day and Night, how the Shadow of the Moon's Body will cover some part of the Earth, and thereby shew you, that to the Inhabitants of that part of the Earth there will be a *Solar Eclipse*.

THAT is exceeding Plain and Instructive, said the Lady; I have taken Notice of two or three already, as you have whirled the Earth and Moon round the Sun. But pray for what other End do you thus turn it now?

ONLY to bring it to shew you the *Autumnal Equinox*, said I, Madam! and then you will plainly see the Reason of the *Equality of Days* and *Nights* all over the Earth, when ſhe is in that Poſition.

O! SIR, ſaid ſhe, I thank you; this explains the Figure you drew for me before, by which alone I could not get ſo diſtinct and ſo clear an Idea of the Earth's two Motions, as thus ſhewn me. But *now* I ſee, that as the Earth turns round her *Axis*, juſt one half of the Equator and all Parallels to it, will be on the Light, and the other half in the Dark; and therefore the Days and Nights muſt be every where equal: For I ſee the *Horizon of the Earth's Disk* now lies parallel to the Plane of the *Solſtitial Colure.*

EXCELLENTLY well remembred and expreſſed, ſaid I, Madam. Your Ladyſhip, I ſee, hath ſtudied hard ſince I ſaw you laſt in the Country, and we are now ſure of you for an Aſtronomer.

I DON'T know that, ſaid ſhe, 'tis probable I may never take pains enough to go into the *Calculatory Part*; but I think every one ſhould be deſirous of knowing the Reaſon of theſe common things we

are

are now upon, and which happen to us every Year. But pray, Sir, go on, and stop when the Earth comes to be in *Cancer*.

'Tis now got thither, said I, Madam; and you will observe that the *Horizon of the Disk*, or that Plane which divides the Earth's two Hemispheres, the *Enlightened* from the *Dark* one, is now no longer *parallel* to, but lies at *right Angles* to the *Plane* of the great *Solstitial Colure*: The Earth being now in *Cancer*, the Sun will appear to be in *Capricorn*; and consequently it will be our *Winter Solstice*. And you see plainly, that as I keep turning the Earth round its Axis either way, the entire Northern frigid Zone, or all Parts of the Earth lying with the Artick Circle, are in the Dark Hemisphere; as you see by this little bit of Wafer, which I stick upon the Circumference of that Circle.

Your Ladyship will observe also, that now I remove that *bit of Wafer*, and place it in the Circumference of that Circle which exhibits the *Path of the Vertex of London*, how much *Longer*, in a Diurnal Revolution of the Earth, *that* will be in the *Dark*, than in the *Light*:

Juſt ſuch is the diſproportion of our Days to our Nights at that time; ſcarce a third Part.

I SEE this thing, ſaid ſhe, exceeding plain; and alſo that the Inhabitants of our *North Pole*, if any ſuch there are, have not ſeen the Sun ſince the 12th of *September*.

No, nor can't again, ſaid I, Madam, till the Vernal Equinox; for all this ſix Months they muſt be condemned to perpetual Darkneſs. But pray obſerve, Madam, that as I move the Earth along in its Orbit, 'till it come thither, how the *Nights ſhorten*, and the *Days lengthen*, by Degrees, till they come then to an Equality again on the 10th of *March*; when our Earth being in the firſt of *Libra*, the Sun muſt appear to be in the firſt Degree of *Aries*. And now the *Earth's Axis*, which you ſee always keeps parallel to its ſelf, will come again to be in the Plane of the *Horizon of the Disk*, and conſequently the Equator, and all its *Parallel Paths* will be *biſſected* by that *Horizon* in every Diurnal Revolution of the Earth; or there will be an *univerſal Equinox* all over the Globe.

THIS

THIS, ſaid the Lady, is indeed ſeeing into the very bottom of the Matter, and underſtanding it from its Cauſes and Original. But pray, Sir, turn about your Handle again; and get me our dear Northern Pole out of the Dark, as I ſee it will ſoon be, and then I hope it will enjoy the Benefit of ſix Months cheering Day, as it hath had a melancholy half Year's Darkneſs.

THAT it will, Madam, ſaid I; and now you will obſerve with pleaſure, how the *Days Encreaſe*, and the *Nights Decreaſe*, as the Earth moves on towards *Capricorn*, where now I will ſtop it; while you obſerve that all the *Polar Circle* is got into the enlightned Hemiſphere; as alſo above two parts in three of the *Path of London* (*b Lmf*) in *Fig.* V. and therefore now our Days are at *Longeſt*, this is our *Summer Solſtice*, or *Midſummer*.

YES, ſaid ſhe, I ſee it, and underſtand it perfectly well: But I ſee withal, that our Days, now at their greateſt extent, are going to ſhorten again, which I will bear as long as I can, that is, till you

you wheel the Earth about again into *Aries:* But then, if you please, we will leave off, having attended upon the Earth in one entire Revolution round the Sun; and most demonstratively and delightfully seen, how thereby all the *Phænomena* of the different Seasons of the Year, and the Varieties and Vicissitudes of Night and Day are solved and accounted for.

Pray when you see Mr. *Rowley*, thank him from me, for this most noble and intellectual Entertainment.

Claudiani Epigr. xiii. *In Sphæram Archimedis,*

Jupiter in parvo cum cerneret Æthera Vitro
Risit, & ad Superos talia dicta dedit:
Huccinè mortalis progressa Potentia Curæ?
Jam Meus in fragili luditur Orbe labor!
Jura Poli, rerumq; fidem, Legesq; Deorum
Ecce Syracusius transtulit arte Senex!

Inclusus variis famulatur Spiritus Astris,
Et Vivum certis Motibus urget Opus!
Percurrit proprium Mentitus Signifer Annum
Et simulata novo Cynthia mense redit!

Jamq; suum volvens audax Industria Mundum
Gaudet, & humanâ sidera mente regit,
Quid

Quid falso insontem tonitrue Salmonea (*a*)
(*miror.*
Æmula Naturæ parva reperta Manus.

Thus imitated and applied to Mr. *Rowley*'s ORRERY.

When lately Jove the ORRERY *survey'd,*
He smiling thus to Gods in Council said;
How shall we stint presuming Mortals Pow'r?
The Syracusian Sage did, once before,
The heavenly Motions shew in Spheres of
(*Glass,*
And the Erratick Orbs and Stars express:
But his Machine by one fixt Pow'r and
(*Weight,*
Mov'd, and was govern'd, as we are, by Fate.
While the bold Rowley *in his Orrery*
Keeps his first Pow'r, just like his Genius,
(*free:*
He knows the secret Springs; and can im-
(*part*
Laws to the whole, and to each single part;
His daring Hand, or brings or hinders Fate,
Makes Mercury *fly, or* Saturn *walk in State:*
He

(*a*) *Salmoneus* King of *Elis*, by driving a Chariot over a Brass-bridge, dared to imitate Thunder, for which *Jove* slew him with a Thunderbolt; for thus *Virgil*, *Æn.* 6. speaks of him,

Vidi & Crudeles dantem Salmonea pænas,
Dum Flammas Jovis, & sonitus imitatur Olympi.

He makes the Earth thro' silver Zodiac *run*
Justly obsequious to the Golden Sun:
While the bright Moon shining with bor-
(row'd Light,
Marks out the Months, and rules the Sable
(Night.
And all obedient to his sole Command,
Turn round their Axes, as he turns his
(Hand:
Their Phases *and their* Aspects *all display,*
And at his beck, exhibit Night or Day:
He makes Eclipses as he will appear,
For any past, present, or future Year;
Shews their true Cause, and roots out
(vulgar fear.
Guiltless Salmoneus *at your Suit I slew,*
Shall I to please you take off Rowley *too?*
O! no! all cried; the glorious Artist
(spare;
Transplant him hither, and make him a Star.

This famous Sphere of *Archimedes* is mention'd by *Cicero* and by *Ovid:* and the *former* saith, that it shewed the Motion of the Sun, Moon, and Planets. *Pliny* tells us, that *Atlas* and *Anaximander*, both made such a Sphere; as *Diogenes Laertius* saith *Musæus* also did. *Sextus Empiricus* saith it was made of Wood; and *Cælus Rhodiginus*, that it was of Brass.

F I N I S.

Printed in the United States
By Bookmasters